STARFLIGHT
AND OTHER IMPROBABILITIES

BOOKS BY BEN BOVA

Milky Way Galaxy:
Man's Exploration of the Stars
Star Watchman
Uses of Space
Weathermakers
Out of the Sun
The Dueling Machine
In Quest of Quasars:
An Introduction to Stars and Starlike Objects
Escape
Planets, Life and LGM
Exiled from Earth
THX 1138 (with George McLucas)
Fourth State of Matter:
Plasma Dynamics and Tomorrow's Technology
The Amazing Laser
Starlike Objects
Starflight and Other Improbabilities

STARFLIGHT AND OTHER IMPROBABILITIES

by BEN BOVA

THE WESTMINSTER PRESS
Philadelphia

Book Design by Patricia Pennington

PUBLISHED BY THE WESTMINSTER PRESS®
PHILADELPHIA, PENNSYLVANIA

PRINTED IN THE UNITED STATES OF AMERICA

Library of Congress Cataloging in Publication Data

Bova, Benjamin.
Starflight and other improbabilities.

SUMMARY: Speculative essays about man's future activities in space based on his present knowledge of astronomy and astrophysics.
CONTENTS: Starflight.—Where is everybody?—Yardsticks in space. [etc.]
1. Astronomy—Juvenile literature. [1. Astronomy] I. Title.
QB46.B773 520 72–7332
ISBN 0–664–32520–3

To Arthur C. Clarke,
who got us started

Contents

1.
Starflight

IMAGINE A TELEVISION TALK SHOW, a panel discussion, where the moderator or "anchor man" is a science-fiction writer.

He's sitting at a large round table of gleaming dark wood with touches of shining metal around the edges. Sitting at the other chairs around the table are experts from every field of science. Farther away from the table are the cameras and lights and technicians.

The science-fiction writer starts the discussion:

"In the old days—say, fifty years ago—science-fiction writers had a much easier job than we have now. For example, when Edgar Rice Burroughs wanted a strange and marvelous setting in his stories, he simply placed his characters on Mars or Venus or even in the center of the Earth. Nobody could tell him he was wrong."

"I thought Burroughs wrote about Tarzan," says the astronomer, two seats down from the writer.

Nodding, the writer answers, "He did. But he also wrote terrific adventure stories that were set on Mars and Venus and in Pellucidar, the land *inside* the Earth."

"A hollow Earth?" the geophysicist grumbles. "Nonsense."

"That's just the trouble," the writer says. "Today I can't get away with such stuff. People know that the Earth isn't hollow. They know that Venus is a red-hot desert underneath its clouds, where the ground temperature's hot enough to melt aluminum and the air is an unbreathable soup . . ."

The astronomer agrees. "Terrible place."

"And Mars," the writer continues, with something like a tear gleaming in the corner of one eye, "good old Mars—we've *photographed* it. No canals, no cities, no Martian princesses. Nothing but craters and desert."

"Well, what's your problem?" the physicist asks sharply. She's sitting across the table from the writer. "I should think you'd be glad to get all this accurate information. Now you can write stories that have real, accurate backgrounds."

"Yes, that's true," the writer admits. "And sometimes I do write such stories. But where do I turn to when I want an absolutely wild and wonderful planet? I can't tell my readers that Mars has cities of glass surrounded by fields of purple trees. I can't tell them that Venus is populated by sea serpents that sing underwater . . ."

"Why on Earth would you want to tell anyone such nonsense?" the medical doctor asks, looking startled.

"Because sometimes I just want to write a story that

goes way beyond the everyday world. Sometimes I need a strange and wonderful landscape for a story. And there aren't any . . . not in our solar system."

The entire panel of experts falls silent. Finally the writer says, "So where can a science fiction writer turn if he wants to create a bizarre world with strange-looking alien creatures? He's got to go outside our solar system. He's got to set his story on a planet that circles a different star.

"And that's the purpose of our discussion, ladies and gentlemen. You're going to tell me how to get to the stars."

"Impossible!" the physicist snaps, with a shake of her head. "You just can't get to the stars. Not really. A rocket would have to reach a burnout speed of nearly 58,000 kilometers per hour to escape from the Sun's gravitational influence. The best we've been able to achieve to date is somewhere around 40,000 kilometers per hour, for the rockets that have gone to the Moon and the planets."

"We could improve the rockets," the writer says.

"Please let me finish," says the physicist. "Even if the rocket could keep going at its top speed of 58,000 kilometers per hour—which it won't, once the rocket engines shut down—it would take 80,000 years to reach Alpha Centauri, the next nearest star. Actually, figuring that the rocket will slow down and coast almost all the way, it will take more like a million years for a Saturn type of rocket to get to Alpha Centauri."

The astronomer adds, "And that's the star that's closest to our solar system. It's only 4.3 light-years away. Even if you could travel at the speed of light—which is 300,000 kilometers *per second*—it would take 4.3 years to get there."

"A light-year is a measure of distance, isn't it?" the science-fiction writer asks.

"Precisely," answers the astronomer. "Light travels at 300,000 kilometers per second. There are about 31.5 million seconds in a year. Therefore light can travel nearly nine and a half trillion kilometers in a year. This distance is used as a unit of measurement in astronomy and called a light-year."

"And the nearest star is 4.3 light-years away," adds the physicist, somewhat smugly.

"Which means it would take you 4.3 years to get there even if you traveled at the speed of light," the astronomer says.

Undaunted, the science-fiction writer asks, "Well, what about hyperspace? I can send my starship into super-light-speed overdrive and leave this continuum for a different dimensional coordinate system. In hyperspace I can travel as fast as I want to. I can get my hero to Aldebaran VI just in time to win the battle . . . or just late enough to allow the heroine to have been abducted by the tarantula people . . ."

The physicist throws her hands up and nearly screams. "Gibberish! Absolute nonsense. You just invented hyperspace as a convenience."

NASA

Man's first star probe lifted off from Cape Kennedy in March, 1972. The Pioneer 10 *probe was aimed at the planet Jupiter, not at the stars. But after it has flown past the largest planet in our solar system, the spacecraft will receive a "boost" from Jupiter's gravity that will fling it outward, away from the solar system. In thousands or millions of years, it might reach the vicinity of another star.*

"Not me," the writer counters. "The mathematicians did. And anyway, since you can't prove that I'm wrong, the idea is still useful. Science fiction can always use an idea that hasn't been proved wrong. Right?"

Muttering something about the conservation of energy, the physicist sinks back into her chair.

"She's got good reason to be upset," the astronomer says. "This hyperspace business is merely a mathematical trick. It's got nothing to do with the real world. Nothing in the universe has been observed to travel faster than light. Chances are that nothing ever will. No . . . if you want to go to Alpha Centauri, it's going to take you at least 4.3 years. The speed of light is a barrier that apparently can't be broken through."

The science-fiction writer frowns. "The 'light barrier' sounds a lot like the 'sound barrier' that the aircraft engineers broke through about thirty years ago."

All the scientists groan in protest. The engineer, waving his slide rule as he speaks, explains that everyone knew that the speed of sound was not a fundamental barrier.

"Bullets, meteors, shock waves . . . there were plenty of things that we knew traveled faster than sound. The problem was to build an airplane that could do it without breaking up because of the stresses it flew into right around Mach 1—sonic speed."

"But *nothing* goes faster than light," the physicist resumes. "Nothing in the whole universe."

"Isn't there something called tachyons?" the writer

asks, his voice somewhat subdued. "Don't they go faster than light?"

"Tachyons are completely theoretical," the physicist explains. "A few physicists have been playing with the idea that there are particles somewhere in the universe that go faster than light, and they called these particles tachyons. But as far as we know, they don't exist. In fact, according to the theories, if tachyons did exist we could never see them or use them . . . so it doesn't really matter whether they exist or not!"

"This is beginning to sound awfully gloomy." The writer leans back in his chair, sulking.

"You asked us and we're telling you," the physicist insists. "Look. At the speed of light, the basic relationships between matter and energy begin to do strange things. For example, we've built powerful machines called synchrotrons. In them we can speed up the motion of an electron until it's going at better than 95 percent of the speed of light. At those speeds, when we add more energy to the electron, it doesn't go any faster—it gets heavier! Its mass grows. No matter how much energy we pump into the electron, it never reaches light speed. The energy changes into mass."

The writer looks ready to slide under the table and forget the whole affair.

But from a quiet part of the table the mathematician says, in a thin but clear voice, "Although the physicist and engineer can't tell you how to build a ship that goes as fast as light, I can show you what would happen if such a ship existed."

"You can?" The writer perks up.

"Certainly," replies the mathematician. "Please remember that a mathematician named Newton showed three hundred years ago that an artificial satellite could be established in orbit. Mathematicians can tell you what the stars will look like a billion years from now, or what interactions a mu-meson will undergo in its first millionth of a second of lifetime, or . . ."

"Okay, okay," says the science-fiction writer. "What about starships?"

"First," the mathematician says, "it's not necessary to travel exactly at the speed of light. If the ship could get to within a few percent of light speed, then time would begin to change aboard the ship.

"This all stems from Einstein's theory of relativity," he adds. "Although most people claim Einstein was a physicist, he was really quite a mathematician as well."

"Spare us the commercial," the medical doctor mumbles.

Sniffing slightly, the mathematician goes on, "The physicist told you that strange things begin to happen to matter and energy when you get close to light speed. Well, strange things happen to time, as well.

"The mathematics of relativity," he explains, "show that if a ship were to approach the speed of light, time aboard the ship would slow down. A clock aboard the ship would tick slower and slower as the ship's speed got closer and closer to the speed of light. Everything aboard the ship, the human crew included, would slow down with respect to time on Earth. But aboard the ship itself, nothing would seem to change. Everything would seem quite normal, even though years of time might pass on Earth before a second elapses on the ship.

"This is the basis of the famous 'twin paradox' of relativity. If one twin brother stayed on Earth while the other flew to a star at nearly the speed of light, when the flying twin returned to Earth, he would be younger than the brother he left behind.

"The German mathematician Eugen Sänger once gave the following example: A ship flying at more than 90 percent of the speed of light travels 1,000 light-years to Polaris, the North Star. Ignoring such details as the time spent accelerating to top speed and decelerating to landing speed again, the ship could make the flight to Polaris and back to Earth in a *subjective* time of 20 years. That is, to the crew on board the ship, only 20 years will have passed. But when they return to Earth, our planet will be 2,000 years older than when they left!"

"That's wild," says the science-fiction writer, looking a little groggy at this point.

The mathematician nods happily. "So you see, if we could travel at speeds close to the speed of light, we could reach the stars. There's no need to break the so-called 'light barrier' to get to the stars."

"Not to all the stars," says thc astronomer. "Just to a handful of stars, the nearest ones. Even at light speed, the stars are too far away."

The mathematician disagrees. "Come now. Sänger showed you could fly across the entire known universe in a subjective time of only 40 years, if you fly at 99 percent of the speed of light."

"And return to an Earth that's billions of years older than when you left it," the astronomer retorts. "Who would go on such a venture? How could you know that the Earth would still exist after such a time?"

"Wait . . . wait . . ." The writer puts an end to their argument before it can go any farther. "If it's mathematically possible to cover such distances, could we really build ships to do the job? I mean, sticking to these ideas that there is a 'light barrier' and that nothing can go faster than light, can we someday build starships that will go at least *close to* the speed of light?"

"You might not have to," says the engineer. "There's always the possibility of an interstellar ark. You know, a huge ship with a completely self-sufficient colony aboard. They'd sail out toward the stars at speeds not much more than solar escape velocity—that 58,000 kilometers per hour we were talking about a few minutes ago."

"But that would take thousands of years . . . millions . . ."

The engineer shrugs. "Sure, it would take generations and generations. People would be born aboard the ship, live out their lives, and die. Their great-great-many-greats-grandchildren would eventually get to the star they were aiming for. But that would be the simplest kind of ship to build. Awfully big, of course—like a moving city. But it could be built. I think."

The psychiatrist, who's been silent up to now, says, "I doubt that normal, well-adjusted human beings would ever embark on such a journey. How could they, in good conscience? They'd be dooming many generations of their offspring to live and die aboard the ark. How do they know that the children who finally reach their destination star will want to live there?"

"Or," the astronomer adds, with a twinkle in his eye, "that another group in a faster ship hasn't beaten them to it?"

"Even leaving that possibility aside," the psychiatrist continues, "no group of human beings who could be considered to be normal would ever contemplate such a mission. Why, they would have to be a group of exiles. Or religious fanatics."

"Like the Pilgrims or Quakers?" somebody asks.

The engineer says, "I'm assuming that the rocket engines aboard the ark will be based on nuclear fusion. You know, the hydrogen fusion process, such as the Sun and stars use. Hydrogen atoms come together to make a helium atom, and release energy."

"No one's built a fusion rocket," the physicist points out. "In fact, even the fission rockets—the kind that use uranium or plutonium, where the atoms are split to release energy—they're still in the testing stage. Nobody's flown one. And the only way we've been able to release fusion energy here on Earth is in hydrogen bombs."

"I know," the engineer admits. "But progress in fusion research has been very encouraging over the past few years. I think we can safely agree that fusion power will be available before the end of this century."

"Perhaps," the physicist says reluctantly.

"Fusion rockets will make tremendous propulsion systems," the engineer says glowingly.

The engineer goes on to explain about a study undertaken by Dwain F. Spencer and Leonard D. Jaffe at the California Institute of Technology's Jet Propulsion Laboratory. "Spencer and Jaffe assumed that fusion rockets could be built, and then they tried to design a starship that uses fusion power. The ship they came up with—on paper—had five stages, each one powered by fusion rockets. It can make a round-trip flight to Alpha Centauri in a total elapsed time of 29 years. The ship would accelerate at 32 feet per second, every second, for several months. This is the same force that we feel here on Earth due to our planet's gravity. So, during the ship's acceleration period, the crew would feel 1 *g*, their normal Earth weight.

"After several months of this acceleration, the ship would be traveling at a relativistic speed—fast enough for time effects to come into play. It would then shut down its engines and coast the rest of the way to Alpha Centauri. The same procedure would be followed for the return trip: a few months of 1 *g* acceleration, then coasting flight back to Earth.

"The 29 years would seem slightly shorter to the

ship's crew," the engineer says, "because of the relativistic time-dilation effect."

"And that's using power that we know we can harness," the science-fiction writer adds excitedly. "Why, maybe by the end of the century we could reach Alpha Centauri! People alive today might make the trip!"

"Excuse me," says the astronomer. "Have any of you heard of the Bussard interstellar ramjet?

"R. W. Bussard was a physicist at the Los Alamos Scientific Laboratory when he thought of the interstellar ramjet idea," the astronomer explains.

"Bussard realized that one of the main drawbacks to any rocket engine is that it must carry all of its propellant with it. Spencer and Jaffe's five-stage fusion rocket, for example, must be more than 90 percent hydrogen propellant—allowing very little payload for such a huge vehicle. The rocket must also spend a considerable amount of its energy just lifting its own propellant mass. The situation becomes a vicious circle. As long as you must carry all the rocket's propellant along with you, any increase in speed must be paid for by more propellant mass. When you're considering flight at close to the speed of light, this becomes a serious obstacle. It poses a fundamental limitation on the amount of energy you can get out of the fusion rocket.

"But suppose the interstellar ship didn't have to carry any fuel at all? It could carry much more payload. And its range would be unlimited—it could go anywhere, at close to light speed, as long as it could somehow find propellant to feed to its engines.

"Interstellar space is filled with propellant for a hydrogen fusion rocket—hydrogen gas. There is enough hydrogen gas floating freely among the stars to build billions of new stars. This is an enormous supply of propellant.

"However," the astronomer admits, "when I use the word *filled* I'm being a little overly dramatic. The hydrogen gas is spread very thinly through most of interstellar space . . . no more than a few atoms per cubic centimeter. By contrast, there are more than 10^{19} atoms per cubic centimeter in the air we're breathing. That's ten million trillion atoms in the space of a sugar cube. Out among the stars, there are fewer than ten atoms per cubic centimeter.

"Bussard calculated that the ramjet will need a tremendously large scoop to funnel in a continuous supply of hydrogen for the fusion rocket engines. For a ship with a payload of 1,000 tons—about the size of a reasonable schooner—a funnel some 2,000 kilometers in radius would be needed."

The mathematician smiled. "I'm tempted to say that such a scoop would be *astronomically* big."

"Yes," the engineer says, "but there's plenty of open space out there."

"And the scoop needn't be solid material," the physicist adds. "If you could ionize the hydrogen with laser beams, so that the atoms are broken up into electrically charged ions, then the scoop could be nothing more than an immense magnetic field—it would funnel in the electrified ions quite nicely.

"Such a ship," the astronomer goes on, "can reach the nearest stars in a few years—of ship time, that is. The center of the Milky Way would be only about 20 years away, and the great spiral galaxy in Andromeda could be reached in about 30 years. Of course, the elapsed time on Earth would be thousands, even millions of years."

"Even forgetting that for a moment," the science-fiction writer asks, "don't you think the crew's going to get bored? Spending 20 or 30 years traveling isn't going to be much fun. And they'll be getting older . . ."

A polite cough from the other side of the table turns everyone's head toward the biochemist.

"As long as we're stretching things," he says, "we might as well consider the possibility of letting the crew sleep for almost the entire flight—slowing down their metabolism so that they don't age much at all."

"Suspended animation?" the writer asks.

With a slightly uncomfortable look, the biochemist replies, "You could call it something like that, I suppose. I'm sure that by the time we're ready to tackle the stars, a technique will have been found to freeze a human being indefinitely. You could freeze the crew shortly after takeoff and have them awakened automatically when they reach their destination. They won't age while they're hibernating."

"This is the idea of freezing them at cryogenic temperatures, isn't it?" the medical doctor asks.

Nodding, the biochemist says, "Yes. Temperatures

close to absolute zero. More than 400 degrees below zero on the Fahrenheit thermometer."

"That simply can't be done," the doctor says firmly.

"Not now," the biochemist agrees. "But by the end of this century, we might have learned how to quick-freeze live human beings without damaging their cells."

The doctor looks unconvinced and shakes his head.

"I must point out," the psychiatrist says, "that you still have the basic problem of motivation on your hands. Who would want to leave the Earth, knowing that he would return to a world that's several thousand years older than the one he left?"

"It would be a one-way trip, wouldn't it?" the writer muses. "Even if the crew comes back to Earth, it won't be the same world that they left. It'll be like Columbus returning to Spain during the time of Napoleon."

"Or Leif Ericson coming back to Scandinavia next week."

"The crew members will want to bring their families with them," the writer points out. "They'll have to."

"Nothing man has ever done comes even close to such an experience," the psychiatrist says.

"Oh, I'm not so sure about that," objects the anthropologist. He has been sitting next to the psychiatrist, listening interestedly and smoking a pipe through the whole discussion.

Now he says, "The Polynesian peoples settled the islands of the Pacific on a somewhat similar basis. They started in one corner of the Pacific and expanded throughout most of the islands in the central regions of that ocean. And they did it on a somewhat haphazard basis—a mixture of deliberate emigrations into unknown territory plus accidental landings on new islands when ships were blown off course by storms."

"That's hardly . . ."

"Now listen," the anthropologist insists quietly. "The Polynesians ventured out across the broad Pacific in outrigger canoes. Their travels must have seemed as dark and dangerous to them as interstellar space seems to us. They left their homes behind—purposely, in the case of the emigrants. Usually, when they were forced to emigrate because of population pressure or religious differences, they took their whole families along. But they knew they'd never return to their original islands again. That's how Hawaii was first settled, and most of the other islands of the central Pacific."

"That *is* somewhat similar to starflight," the psychiatrist agrees.

"So we can reach the stars after all," the science-fiction writer says. "It's not fundamentally impossible."

"It won't be simple," the engineer insists.

"Yes, but imagine a time when we can travel with interstellar ramjets from star to star."

"You'll never be able to go back to the same place again," the physicist reminds him. "Too much time will have elapsed between one visit and the next."

"I understand," the writer answers. "But consider it: Starship crews would be forced to think ahead in terms of centuries. They'd never know what would be coming up next, what the next world would hold for them. What an age for adventure!"

The mathematician chuckles. "And if a star traveler should deposit a few dollars in a savings account, then come back several centuries later, what an age for compound interest!"

The science-fiction writer turns his beaming face to the panel of experts and thanks each one of them in turn.

"You have certainly answered my problem. I can now write about interstellar ramjets, where the crews are frozen during the travel time from one star to the next. Why—the crew members will become virtually immortal! Who needs Mars? The rest of the universe is going to be much more exciting!"

2.
Where Is Everybody?

SCIENTISTS AND SCIENCE-FICTION WRITERS are always asking themselves questions.

For example: We saw in the first chapter of this book that starflight isn't so improbable after all. We can foresee an era—perhaps before the end of this century—when we'll have the necessary know-how to build starships. All right, now suppose there are other intelligent races out in space, living in other solar systems. Some of them must have reached the point where they already have starships. Why haven't they visited us?

In other words: Where is everybody?

Enrico Fermi, the late Nobel laureate physicist, once asked this same question.

His reasoning was straightforward: The universe is vast. The Milky Way galaxy alone contains some 100 billion stars. And there are billions and billions of other galaxies, each packed with billions of stars. By mere blind chance alone, there must be literally billions of planetary systems harboring intelligent life. It seems incomprehensible that Earth is the *only* place in this enormous universe where intelligent life exists.

Now, astronomers have deduced that our solar system is considerably younger than most of the universe. Studies of the Sun and of the rocks of the Earth, Moon, and meteorites show that our solar system is roughly 5 billion years old. There are much older stars, and astronomical evidence points to the conclusion that the universe, as a whole, is at least 10 billion years old.

(Incidentally, there are much younger stars, too. New stars are constantly being formed out of the loose interstellar hydrogen gas we talked about in Chapter 1.)

If much of the universe is older than we are, then the chances are strong that there are intelligent races out there who are much advanced over us. Starflight should be a relatively simple matter for them.

So again: Where is everybody?

Fermi wondered why—if all this is true—no alien races have established contact with us. Or is the fact that we have *not* received interstellar visitors proof that *no* intelligent life forms exist in space?

Contact needn't be in the form of visitors in a starship, of course. It could be a message coming in on radio or light waves. Radio astronomers have tried to find such signals. Once they even thought they had them.

In 1960, radio astronomers at the National Radio Astronomy Observatory in Green Bank, West Virginia, studied the radio emissions coming from two relatively

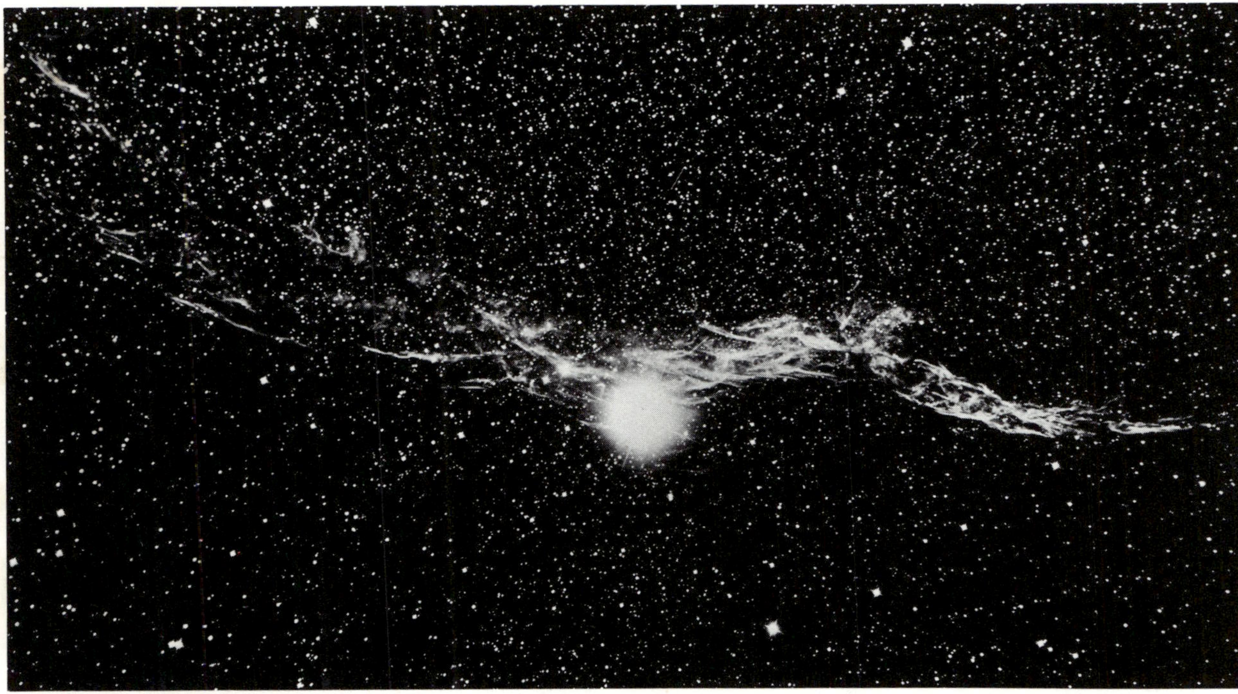

Hale Observatories

The Veil nebula, in the constellation Cygnus (the Swan). Each dot is a star, one of the hundred billion stars in the Milky Way galaxy. The bright spot in the center is the star Deneb, which is closer than the background stars and so is out of focus. The nebulous wisps are clouds of hydrogen, hot enough to glow, probably the remains of an ancient stellar explosion.

nearby stars: Tau Ceti and Epsilon Eridani. Both are about 10 light-years from Earth, and both were thought to be the type of star that is long-lived and stable enough to allow life to develop. There was no evidence of planets around either star, but that doesn't mean that there are none. At such distances from us, planets are too small and dim to detect.

The study was called Project Ozma, named after the queen of L. Frank Baum's fictional land of Oz.

Between May and June of 1960, for a total of 150 hours, the 85-foot radio telescope at Green Bank listened for signals of intelligent life from the two stars. Nothing was heard except completely natural radio "noise."

That doesn't mean there is no intelligent life in space, of course. For one thing, Ozma concentrated on two stars out of billions. Only 150 hours of listening time could be squeezed out of the radio telescope's busy scientific schedule. And the radio telescope wasn't powerful enough, really, to pick up anything but the strongest kind of signals, beamed directly at us.

Most astronomers have concluded that they need much more powerful and sensitive radio equipment before they can hope to make contact with another intelligent race. The problem is something like what we would face if we were on a distant star and trying to detect Earth's commercial television and radio broadcasts. For one thing, since the very first radio broadcasts were made only about 50 years ago, by Marconi, our radio signals have not reached out into space any farther than some 50 light-years. If alien astronomers are now looking toward Earth from farther away than 50 light-years, they won't get any radio signals from us, no matter how powerful and sensitive their equipment.

But in the summer of 1967, radio astronomers at Cambridge University began to detect strange, pulsed radio signals from the heavens.

Nothing like this had ever been found before. The radio signals came in pulses, bursts that were only 10 to 20 thousandths of a second long and came every 1.33730113 seconds. The precise timing of the pulses was as accurate as any chronometer on Earth!

The precision of these pulses tempted astronomers to wonder if the signals were coming from an intelligent civilization. For a few weeks they debated the "LGM theory" as one way of explaining the radio pulses. LGM stood for "little green men."

But by the end of 1967, completely natural reasons for the *pulsars* had been discovered, and the LGM theory was put to rest.

As you can see, establishing contact with another intelligent race by radio, or laser beam, or some other signaling device won't be simple. First, we must decide whether the signal we're receiving is natural or "man-made." Then we will have to decode the signal to make some sense out of it. Only after the message is decoded and sensible can we be really sure that it is a contact from another intelligent race and not some unusual but completely natural phenomenon.

It would be much more dramatic, of course, if a starship flashed across our skies and landed a few ambassadors on the surface of our Earth. Let's take a look at some of the reasons why this hasn't happened. The reasoning we use could apply almost equally to radio or other forms of signaling, as well as to actual visits by alien astronauts.

Two types of reasons are usually given for our lack of interstellar visitors. The first is the "grain of sand" argument. The second is the "postage stamp" analogy.

The first argument uses a poetic metaphor to make its point:

A man can walk across a very large beach without much difficulty. He can chart its shoreline and extent, its contours and headlands. But—*can he inspect every grain of sand on the beach?*

In other words, even assuming that an advanced race could develop starflight, could they explore every one of the Milky Way's 100 billion stars in an effort to find other intelligent races?

Stated this way, the prospects for interstellar contact sound dim indeed. But let's examine this argument a little more closely. Basically, it involves two facets: the ability to achieve interstellar flight, and the ability to investigate very large numbers of stars.

We saw in Chapter 1 that starflight might be possible for us by the end of this century. For a race that's technologically far ahead of us, starflight should be very likely.

But, even in the best starships imaginable, what about examining every grain of sand?

Well, it's not going to be necessary to examine every star in the Milky Way in a search for intelligent life.

Many stars are just plain inhospitable, and it is hard to see how intelligent life could have evolved near them. The brightest and biggest stars are relatively young; yet we know that it took billions of years for life to evolve on Earth, and even longer for intelligence to arise. So stars that are younger than 5 billion years could be ruled out in any search for intelligent life. Most of the stars in the Milky Way are small, faint dwarfs. Although they are very long-lived, they are so cool that the chances are slim that planets might be orbiting them and be warm enough for life to arise. If Earth, for example, were circling around such a cool red dwarf star at the same distance it orbits the Sun, our planet would be a frozen ice ball—life would be impossible.

So an intelligent race that is scouting through the Milky Way for other intelligent races could rule out a tremendous number of the stars as unsuitable for the development of intelligent life. That still leaves a monumental task.

With the use of sensitive detection instruments such as radio receivers and optical telescopes, starfaring aliens could detect life on Earth from a distance of at least 50 light-years. So it seems possible, at least, that an intelligent race could indeed find our particular grain of sand, given enough energy, time, and purpose.

Hale Observatories

The Cone nebula, in the constellation Monoceros (the Unicorn). One of the scenes of overwhelming beauty to be found in interstellar space. Here, within clouds of dark dust, new stars are probably being formed.

The question of *purpose* brings us to the "postage stamp" analogy:

Both astronomers and anthropologists have often painted the following picture: Consider the history of planet Earth. Let the height of the Empire State Building represent the planet's 5 billion years of existence. Man's measly million years or so of life can then be represented by a one-foot ruler standing at the very top of the building. The thickness of a dime placed atop the ruler represents the length of man's entire history of civilization—about 10 thousand years. And at the very top of the whole wobbly structure we glue a postage stamp. This represents the length of time since man has developed modern science, about 300 years.

If other intelligent races exist out among the stars, what are the chances of our meeting a race at exactly our own level of development—that is, within the thickness of the postage stamp? The overwhelming chances are that any intelligent race will either be far below us or far above us. If their starships come to Earth, then obviously they will be above us.

But wait.

Several cosmologically-minded thinkers have come to the conclusion that science and technology may be only a passing phase in the development of an intelligent species. Perhaps only in the first blush of youth does an intelligent race care about exploring the stars.

This kind of reasoning is typified by Sebastian von Hoerner, a German scholar. He states that an intelligent race is bound either to destroy itself or to stagnate within a few hundred or, at best, a few thousand years after reaching the modern Earthly level of technology. In other words, the postage stamp may grow as thick as the dime, but certainly no thicker. We will either blow ourselves up, or forget about technology and return to nature.

Is this a reasonable assumption? Will man destroy himself? Or will we become passive lotus-eaters, allowing most of our billions of population to starve to death while our machines turn to rust?

Let's be optimistic and assume that mankind (or any intelligent race) won't destroy itself. After all, we have had the capacity to unleash nuclear holocaust for a generation now, and have carefully steered away from such a doom. We are faced with pollution and population problems, it's true, but these can also be solved without killing off the whole species of mankind.

But will we become stagnant? Is the technological "state of mind" merely a passing fancy? Anthropologists have amassed some solid evidence that points exactly in the opposite direction.

Even before man was fully human, he was a maker and user of tools. The wheel and the plow were invented about 10 thousand years ago. The so-called modern era of science, dating roughly from Copernicus and Galileo, is not completely different from the eras that preceded it. The technology that we're so justly proud of didn't spring full-blown from the minds of a few brilliant men. It was the product of many generations of effort. Modern society represents not so much a break with the past as an acceleration of past trends, speeded by the gathering forces of technical methods and accumulating scientific knowledge. In short, an intelligent race is apt to be technologically oriented, and *un*likely to give up its technology.

For man without his technology would be like an ape without teeth, a lion without claws, a fish without gills. Technology is the way we deal with the world around us. We simply cannot continue to exist without it. True enough, we have often used our technology unwisely and polluted our environment. But our technology today is fully capable of producing all the energy and food we require, with only minimal pollution and little harm to the environment.

It would seem, then, that the postage stamp atop the Empire State Building is a trick of perspective. Man's technology is very young, but so is man himself. As long as he has been human, he has been a tool wielder. If and when we meet other intelligent races, the chances are that their technologies will be fully as old as they are themselves. So, if an older race comes to visit us, its technology will be superior to ours.

We've now tested two lines of speculation and concluded that (1) an intelligent race could reach us if it wanted to; and (2) once a race develops technology, it is not likely to dispose of it and return to a primitive existence.

But the original question remains unanswered. If intelligent races abound among the stars, why haven't they visited us? Or is man alone in his intelligence and technology?

Imagine a race of intelligent creatures, like human beings, living in their own world. They have developed in isolation from outside contacts, as we have, and have split up into many local cultures. Some have advanced to high civilizations, others remain struggling in a primitive existence. But all of them are members of a fully intelligent species. Suddenly, their world is visited by a vastly superior race. To simplify matters, we'll assume that both races are human in form. They look very much alike; there are some slight differences, but not many.

The first contacts between these two races are friendly enough. Soon, though, it becomes clear that the visitors have measured the natives and found them lacking. The visitors begin taking over the natives' world. Fighting begins. The natives lack the advanced technology of their opponents. Within a few generations the natives cease to exist, except for a few scattered remnants of tribes in the hills and deserts.

The natives have not been merely beaten in a war. They've been virtually extinguished by a stronger culture. Through intermarriage, through susceptibility to new diseases, through an emotional response that can only be described as "culture shock," the natives either die away or are genetically engulfed by the newcomers.

Yerkes Observatory

Star clouds in the Milky Way, looking toward the constellation Cygnus (the Swan). Each pinpoint of light is a star.

This actually happened to the American Indians.

What would happen if a vastly superior race suddenly dropped out of our skies, straightened out our political squabbles, handed us a child's primer of hydrogen fusion reactors, and generally took over the planet? Could our deeply ingrained pride and culture stand such a shock, or would we go into a racial decline?

Look at it another way. Anthropologists are interested in studying man's nearest relatives, the primate apes. A good deal has been learned by observing chimpanzees and gorillas in captivity. But the basic question of why we live in cities while our closest relatives live in trees can be answered only by studying the apes in their natural habitat. This isn't easy, because the key to the scheme is that the animals being studied *must not know they're being watched.* Only by remaining "invisible" can the scientists learn how the apes behave naturally.

Now let's consider the reactions of a highly advanced race that discovers intelligent life on planet Earth. It seems reasonable to suppose that the ethics of such a race will advance together with its technology, even if the ethics advance more slowly. Any race capable of developing starflight, it would seem, should also be intelligent enough and ethical enough to observe a relatively primitive race such as our own without interfering with us.

Why should they contact us? They have far more to learn by keeping us under surveillance. They might well have a "closed door" policy about contacting us, but an "open window" attitude about observing us.

Inevitably, this line of thought brings us to the question of unidentified flying objects—"flying saucers."

While most UFO sightings have turned out to be perfectly natural things such as weather balloons, artificial satellites, high-flying aircraft, etc., there are still thousands of unexplained sightings. And that is just what they are—no more. Unexplained sighting of unidentified objects. Never has there been a single shred of authenticated evidence to show that the UFO's are actually from another world.

Is it likely that intelligent star rovers, who can build ships that cross the galaxy, who are wise enough to observe us without trying to interfere with our affairs—is it likely that these creatures would allow themselves to be spotted in such haphazard fashion? Most of the flying saucer stories are demonstrably silly, and many of the so-called photographs of saucers are clearly faked. If we are being observed by another intelligent race, they're not revealing themselves at all.

Where is everybody?

If we assume that (1) an intelligent race can develop starflight, (2) such a race can detect signs of intelligence at great distances in space, and (3) one or more such races have indeed evolved among the Milky Way's hundred billion stars—then the answer may be this: they may be watching us right now, using us to learn more about this phenomenon called intelligence, waiting for us to reach the maturity necessary before we can join them as brothers and equals.

3.

Yardsticks in Space

MEASUREMENT is the heart of science. In our discussion of starflight so far, we've mentioned the staggering distances between the stars. Now we're going to take a look at what those distances really are and how they're measured.

In astronomy, distance measurements are the foundation upon which rests almost all our knowledge of the universe. For example, an astronomer sees a star in the sky and notes its apparent brightness. But how much of this brightness is due to the star's actual output of light, and how much is due to its distance? A dimmer-appearing star might actually be more luminous than one that looks bright to us, but its greater distance from us may be obscuring its true brightness.

Until a star's true output of light (called its *intrinsic luminosity*) is known, the astronomers can't determine much about its age, evolution, or size.

Another example: How big is the universe? Does it actually have limits, or is it infinite? Distance measurements are vital to such cosmological questions. Before suitable distance measurements could be made, it was impossible to decide whether the so-called spiral nebulas, such as the great spiral in Andromeda, were clouds within our own Milky Way galaxy or distant galaxies in their own right.

How can we measure the distance to a star, or to the farthest reaches of the known universe? We begin with a triangle.

George Washington knew how to measure the distance to an object he couldn't touch. He could measure the width of a river, for instance, with his surveyor's base line and transit telescope.

First he would measure out a base line on one bank of the river. This usually consisted of a chain or cord of precisely known length. Then, from each end of the base line he would set his sight on an object on the other side of the river—a tree, perhaps—right along the water's edge. He would measure the angles formed between each end of the base line and the line to the target object. Thus he could construct on paper a triangle in which he knew the length of the base, and the size of two of the three angles. Simple trigonometry would tell the length of the other two legs of the triangle, and the length of a line drawn from the tree perpendicular to the base line—or the width of the river. That's why surveyors carry well-worn pocket-sized books of trigonometric tables. Or at least they did, until laser transits began to allow them to make their measurements much more simply and precisely.

If Washington had been inclined toward astronomy instead of politics, he could have made accurate measurements of the distances to the Moon and planets. For these problems, astronomers use the triangulation method, but they need much longer base lines—hundreds, even thousands of kilometers long. This is necessary because the distances they're trying to measure are so large.

The longer the base line, the easier it is to make the measurement. Remember, the important measurement is the angle that each end of the base line makes with the target. If the base line is too short, these angles will both seem to be 90 degrees, and they provide no triangulation information at all. The longer the base line, the more the angles vary from 90 degrees, and the easier the distance measurement is.

There are other ways of measuring interplanetary distances, including radar. The importance of these measurements within the solar system is that they produce a precise value for the distance between the Earth and the Sun. This distance is called the Astronomical Unit. Since the Earth orbits the Sun in an elliptical, not a circular, path, our distance from the Sun is not constant. The Astronomical Unit (abbreviated AU) is half of the longest diameter of Earth's elliptical orbit. The most precise measurement to date sets this distance as 149,598,000 kilometers. For our purposes we can say that the AU is about 150 million kilometers or roughly 93 million miles. Precise measurement of the AU is im-

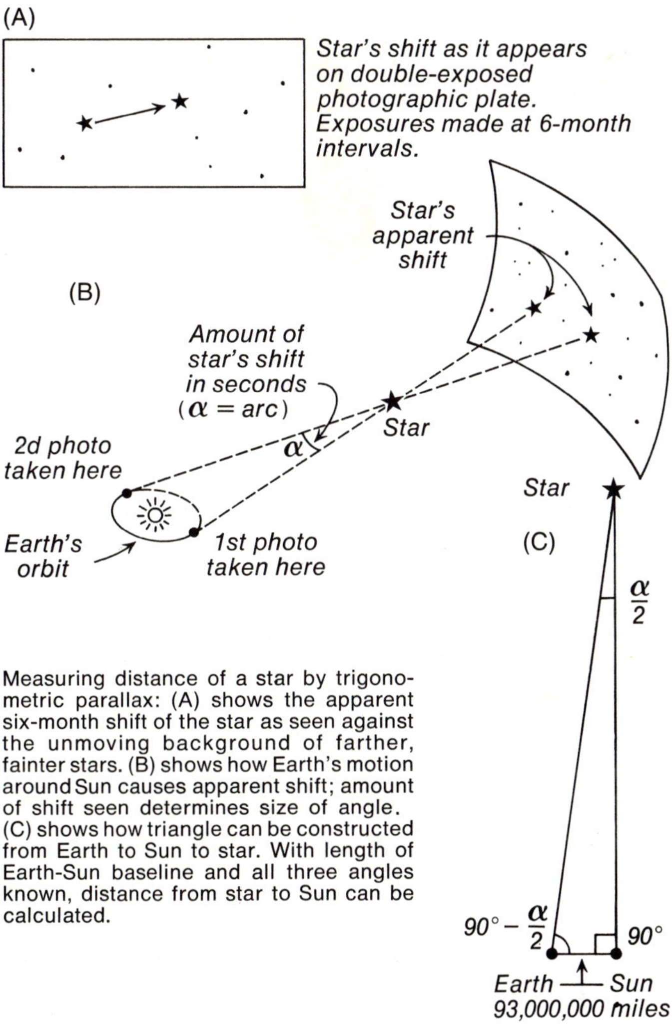

Measuring distance of a star by trigonometric parallax: (A) shows the apparent six-month shift of the star as seen against the unmoving background of farther, fainter stars. (B) shows how Earth's motion around Sun causes apparent shift; amount of shift seen determines size of angle. (C) shows how triangle can be constructed from Earth to Sun to star. With length of Earth-Sun baseline and all three angles known, distance from star to Sun can be calculated.

The trigonometric parallax technique for measuring the distance of a star

portant: this 150-million-kilometer "yardstick" is the basic value for all our distance measurements of the heavens.

Although the possibilities of triangulation measurements for stellar distances had been known since the time of Copernicus in the sixteenth century, it wasn't until 1838 that Friedrich Wilhelm Bessel of Prussia made the first successful measurement. The reason was simple: sheer distance.

No base line on Earth is long enough for stellar distance measurements. But astronomers are ingenious, and they eventually turned to a base line that is literally out of this world: the diameter of the Earth's orbit around the Sun.

But even this astronomical base line is barely long enough.

The technique for stellar triangulation measurements is to observe the star at six-month intervals—that is, from opposite ends of the base line. The target star may show a slight shift of position against the background of farther, dimmer stars around it. This depends on the target star being near enough to show a measurable shift, and the background stars being far enough away to seem fixed in their positions. Many an astronomer has spent a couple of heartbreaking years trying to measure the distance to a star that was just too far away to show any shift at all.

The shift in the star's position is called its *parallactic* shift. You can produce a parallactic shift of sorts for yourself by holding your thumb up at arm's length and squinting at it with one eye at a time. Your thumb will seem to move in relation to the background you see it against. In this case, your base line is the distance between your two eyes. The word *parallax* is of Greek origin, and means "the mutual inclination of two lines forming an angle." Apparently the Greeks really did have a word for everything!

The parallactic shift measured by Bessel was for the star 61 Cygni, which shows a shift of 0.294 second of arc. With sixty minutes to a degree of arc, and sixty seconds to each minute, Bessel's measurement resulted in a triangle that very nearly had two ninety-degree angles, one at each end of the base line. The angle produced by one second of arc is about the width of a twenty-five-cent piece seen from a distance of three miles!

The distance to 61 Cygni is 11.1 light-years. By an odd coincidence, there are almost as many inches in a mile as there are AU's in a light-year. So Bessel's triangle, if he had actually tried to draw it in miniature, would have had a base line of one inch, and legs more than 11 miles long!

Alpha Centauri, the closest star to our solar system, is in our Southern Hemisphere. Its distance wasn't measured until sizable astronomical telescopes were built south of the equator. Alpha Centauri's parallactic shift is 0.76 second of arc—corresponding to a distance of 4.3

Cepheid variable stars are bright enough to be seen in other galaxies. Here, two Cepheids in the Andromeda spiral galaxy are marked. By timing the duration of their pulses, astronomers

Hale Observatories

can estimate the true luminosity of the stars, and so derive their actual distances.

light-years. This is the *largest* parallactic shift any star has shown.

Parallactic shift measurements are time-consuming and tedious, but they're the only direct method of measuring a star's distance. And, as we've seen, the method depends on the precision with which the AU is known. Even so, the trigonometric parallax method works for only a handful of stars. A few hundred stars are near enough for parallactic measurements, the rest of the Milky Way's 100 billion members are too far away—they show no discernible parallactic shifts. We can't measure their distances. Not directly.

But astronomers have other tricks for stars that are more than about 100 light-years away (the approximate limit of the parallactic method).

For example: the Sun is moving through space at a rate of 4.2 AU every year, dragging along the Earth and the rest of the solar system with it. Every year, then, we cover 4.2 AU's of distance. In ten years, we've traveled 42 AU's.

Astronomers have used this type of base line to estimate the distances to large groups of stars. The technique won't work for a single star, because it is moving across the sky too, and its individual motion confuses the measurement. But for a large enough group of stars, we can assume that the individual motions cancel out—for every star moving in one direction, there's a star moving the other way—and the measurement can be made. The average distance from Earth to that group of stars, taken as a whole, can be measured. This is known as the *mean parallax* technique. The trick is to pick out a large sampling of stars that all appear to have the same color and brightness; this indicates that they are all at about the same distance from us. This is true because a star's color is an indication of its temperature —blue stars, for example, are hotter than yellow, and yellow stars are hotter than red. And a star's brightness, as we see it from Earth, is due partly to its distance from us and partly to its temperature. So stars of the same color are probably at about the same temperature. And if both their color and their brightness are the same, they're at roughly equal distances from us.

The mean parallax technique provides a rough value for the average distance to a large group of stars. This distance value is only very approximate for any individual star in the group, but for most of the stars in the Milky Way, this is the best that can be done.

However . . .

There are certain stars in the sky that are actually like signposts, and these can be used to measure distances all across the Milky Way and even out to other galaxies. Like truly good signposts, these stars do their best to attract our attention: they blaze with the luminosity of a thousand Suns, and even pulsate, getting brighter and dimmer, like an advertising sign!

These signposts are certain types of variable stars. There are in the sky many kinds of variable stars—that is, stars whose light output fluctuates rather dramatically. One of the best known is Algol, in the constellation of Perseus, which gets brighter and dimmer every few days. Its name is Arabic, from the same basic root word that our word *ghoul* comes from. The ancients were frightened of a star that didn't shine with a constant light, like a good star should. Algol is actually a double star, one bright and one dim, but so distant that we can see only a single point of light. When the two stars are side by side, as seen from Earth, they show the maximum brightness. When the dim star moves behind the bright one, we see Algol's brightness diminish. And when the bright star "hides" behind the dim one, Algol looks its dimmest to us.

The variable stars that astronomers use as signposts are very different, though. They are the types known as *Cepheid* variables and *RR Lyrae* variables.

RR Lyrae itself is a variable star in the constellation Lyra. All the RR Lyrae stars are strictly telescopic objects; although they glow about a hundred times brighter than the Sun, none of the RR Lyrae type of stars are close enough to the solar system to be seen with the unaided eye.

Delta Cephei, the first Cepheid variable to be discovered, can be seen without a telescope. It's the fourth-brightest star in the constellation Cepheus, between Cassiopeia and Cygnus, the Swan. A much more prominent Cepheid is Polaris, the North Star. But don't expect to see any changes in brightness with your naked eyes. Polaris fluctuates only a tenth of a magnitude every four nights. This is a glaring phenomenon for astronomers, but impossible to detect without precise measuring instruments.

What makes the Cepheids and RR Lyrae variables important is that they fluctuate on an extremely regular timing. Astronomers have concluded that these stars are actually pulsating, swelling up and shrinking down again, and doing it as regularly as clockwork. The time interval between one peak of brightness and the next is precisely the same for any individual star of these two types. The American astronomer Edward Barnard claimed that if all the clocks on Earth stopped, he could keep time by a certain Cepheid he had been studying.

The Cepheids are truly very luminous stars, averaging several thousand times the Sun's luminosity, and so they can be seen over immense distances. Their pulsation periods tend to run from one day to several months. The RR Lyrae variables have shorter pulsation periods: from seven to sixteen hours. Each individual star has its own pulsation period, from which it does not vary.

What has all this to do with distance measurements? If you can determine a star's intrinsic luminosity—the amount of light it is actually giving off—and then compare this luminosity to the star's brightness as seen from Earth, you can determine the star's distance. The difference between the actual luminosity and apparent brightness is caused by the distance. For most stars, the actual luminosity can be determined only after the distance is known. But for the Cepheids and RR Lyrae variables, astronomers have learned to gauge their luminosities independently of their distances—simply by timing their pulses.

In 1910, Henrietta Leavitt of Harvard College Observatory began studying recently obtained photographs of some 1,700 variable stars in the Magellanic Clouds. These two star-filled "Clouds of Magellan" are in the Southern Hemisphere and were first seen by Europeans during the voyages of discovery in the sixteenth century. We know today that the Magellanic Clouds are star clusters that lie outside the Milky Way, about 150,000 light-years from the Sun. They have been called satellites of our Milky Way galaxy.

But in 1910 their distance was not known.

Henrietta Leavitt found that there were many Cepheid-type variables in the Magellanic Clouds. And she discovered that the longer a Cepheid's period of pulsation—the time from one peak of brightness to the next, the brighter the star's apparent magnitude. For example, a Cepheid with a period of 100 days was about 16 times brighter than a 10-day Cepheid.

Two other astronomers seized on this vital piece of information and turned it into the best long-distance measuring technique we have. They were Ejnar Hertzsprung, of Denmark, and Harlow Shapley, of the United States.

They both—independently—made a daring assumption. They assumed that all the Cepheids in the Magellanic Clouds are the same distance from us. Obviously they are not, but since the Clouds as a whole are so far from the Earth, the individual differences in distance from one star in the Clouds to the next is quite small, compared to their overall distance from us.

This is rather like saying everyone in Hawaii is the same distance from San Francisco; it's not quite true, but it's a good-enough approximation.

The importance of the Hertzsprung-Shapley assumption was this: it eliminates the effect of distance on the period-brightness relationship of the Cepheids. It meant that the actual luminosity of the Cepheids varied according to their pulsation periods. In other words, the astronomers could now say that *every* 100-day Cepheid is some 16 times brighter than *every* 10-day Cepheid—anywhere in the universe! Regardless of how bright or dim a Cepheid might appear from Earth, its actual luminosity could be determined by timing its pulsations, and its distance could then be calculated by comparing its true luminosity against its apparent brightness.

The astronomers now had a "yardstick" that quickly allowed them to begin measuring the distances all across the Milky Way galaxy and even out to other galaxies. Shapley led the way and showed for the first time that the solar system is far off from the center of the Milky Way, in one of the galaxy's spiral arms. By the early 1920's, it became apparent that the Milky Way was only one star island out of many. The so-called spi-

JUNE 9, 1950

When a star explodes, it temporarily becomes bright enough to be seen as an individual star, even though it is in a galaxy so distant that no ordinary individual stars can be resolved. By esti-

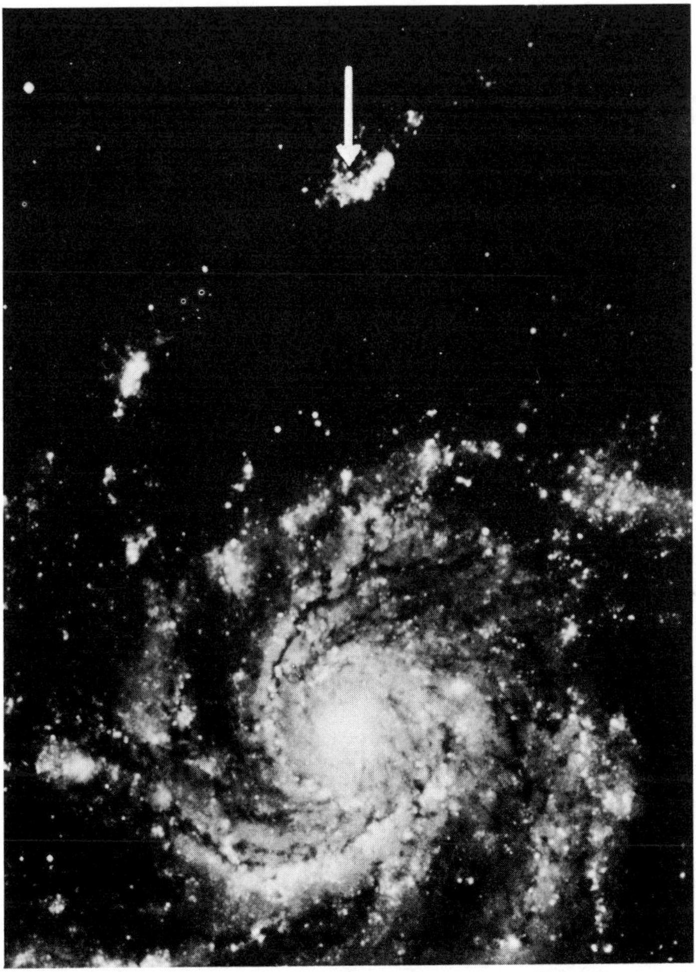

Hale Observatories

FEBRUARY 7, 1951

mating the brightness of this nova explosion in M 101, a distant spiral galaxy, astronomers can determine the distance of the galaxy.

ral and elliptical "nebulas" were other galaxies, other star islands. The nearest ones were close enough so that individual Cepheids could be seen in them, and their distances determined.

It wasn't all downhill sledding, though. It turned out that there are at least two different kinds of Cepheids, with different period-luminosity relationships. So the distance measurements based on the Cepheids have been revised from time to time, sometimes drastically.

For these larger distances, astronomers prefer to use the term *parsec,* rather than light-year. Parsec is a manufactured word, meaning the distance of a star that shows a *par*allax of one *sec*ond of arc. This distance is 3.26 light-years. Actually, as we've already seen, no star has been found that close to us: Alpha Centauri's parallax is 0.76 second of arc.

There's another reason why astronomers prefer to use parsecs rather than light-years. People are always thinking of the term light-year as a measurement of time rather than of distance.

Thanks to the Cepheids, astronomers have determined that the Milky Way galaxy is some 30,000 parsecs in diameter and about 6,000 parsecs thick in its central region. The Sun is roughly 10,000 parsecs out from the center of our galaxy. Beyond the Milky Way, we have the Magellanic Clouds, about 50,000 parsecs distant, and a cluster of 16 other galaxies that, together with the Milky Way, make up the Local Group. The largest of these is the great spiral galaxy in Andromeda, which is about 800,000 parsecs away (2.5 million light-years). The Andromeda spiral is *not* the nearest galaxy to us, as many people have wrongly assumed, but it is the nearest large spiral galaxy, and is about the same size as our own Milky Way.

In 1924, the American astronomer Edwin P. Hubble made the first distance estimates to other galaxies, using Cepheids observed in these galaxies with the 100-inch telescope at the Mount Wilson Observatory. With the distances to a few of the nearest galaxies pinned down fairly firmly, it became possible to estimate the distances of the billions of farther galaxies, by comparing brightnesses. Assuming that galaxies are all of the same order of luminosity (a poor assumption, actually), we can say that a dim galaxy is farther away than a bright one, and most of the difference in brightness is caused by distance.

Then, in 1929, Hubble announced a new way to estimate the distances of the farther galaxies: the *red shift*.

An Austrian mathematical physicist named Christian Doppler suggested in 1842 that the spectrum of a source of light will show a shift toward the red if the source is moving away from the observer at a high-enough speed, and a shift toward the blue if it is moving toward the observer. This is analogous to the change in pitch of a train whistle or fire engine siren as it approaches and then moves away from you. As the source of sound approaches, its pitch goes higher and higher (the sound waves are shifting to higher frequencies); as the sound moves away, the pitch goes lower in frequency.

Except for the very nearest galaxies, the members of our own Local Group, all the billions of galaxies discovered show red shifts. That is, the light coming from these galaxies is shifted toward the red, or low-frequency, end of the color spectrum. This is assumed to be a *Doppler shift,* caused by the motion of the galaxies —they are all flying away from us!

The galaxies of the Local Group show some red shifts and some blue shifts. They are edging around, mostly moving together as a group. But the farther galaxies are all red-shifted. They are all moving away from us. There are huge clusters of galaxies out there, and these clusters are moving more or less in formation, like our own Local Group.

The implication is that the universe as a whole is expanding, like the skin of a balloon, and the galaxies—or galaxy clusters—are all moving away from each other because of this general, universal expansion.

But Hubble found that the farther the galaxy, the larger a red shift it shows. Presumably this means that it is moving away from us at a faster speed. He showed in 1929 that there is a straight-line relationship between a galaxy's observed red shift and its distance from us. This meant that astronomers could estimate the distance of a galaxy merely by measuring its red shift.

It all worked out beautifully well. The bigger the red shift, the faster the galaxy was receding from us, and the farther its distance. The farthest known galaxy is known as 3C 295; its red shift shows that it is receding

QUASI-STELLAR RADIO SOURCES

3C 48

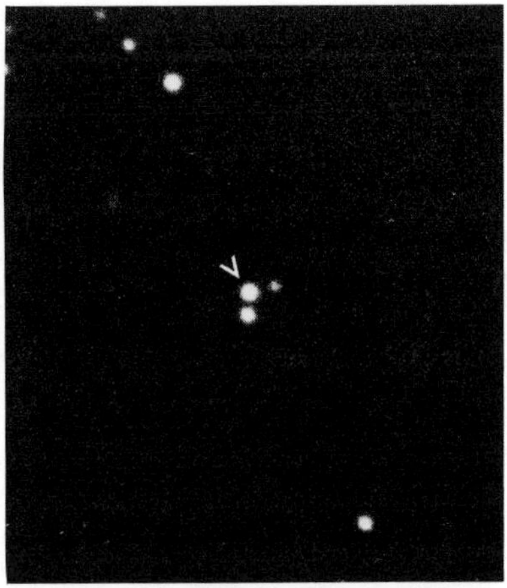

3C 147

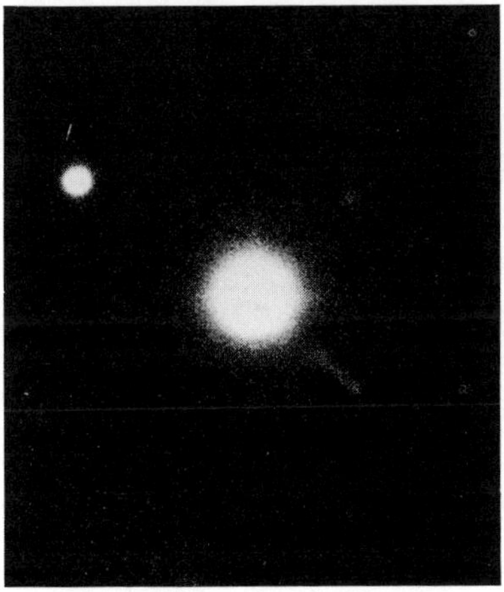

3C 273

Hale Observatories

3C 196

Quasars, as seen by the 200-inch Palomar telescope. The "3C" designation stands for the Third Cambridge Catalogue of astronomical radio sources; the numbers are merely the catalogue listing. While some of the quasars appear to be little more than spots of light, 3C 273 has a faint jet alongside its bright main body. Most of the radio "noise" of 3C 273 comes from the jet.

from us at a speed of 36 percent of the speed of light. This means that it must be on the order of one or two *billion* parsecs away.

Then the quasars were discovered.

These enigmatic powerhouses show red shifts well beyond those of the most distant galaxies. Several quasars have been found with red shifts so huge that they must be traveling at better than 90 percent of the speed of light. This means they must be more than three billion parsecs from us—if their red shifts are truly an indication of their distances.

But the quasars are so puzzling that some astronomers have doubted that their red shifts are truly related to their distances. If the quasars are billions of parsecs away, then with the apparent brightnesses they show, they must be putting out more light and radio energy than a thousand normal galaxies. Yet they appear to be far smaller than an ordinary galaxy! However, if the quasars are much closer, perhaps as close as our neighboring galaxies, then they don't "need" as much energy output to explain their observed brightnesses.

But if the red shifts of the quasars aren't related to their distances, then what about the red shifts of the galaxies? What happens to the idea that the whole universe is expanding, as demonstrated by the red shifts of the galaxies?

If there were some other way to measure the distance to just one quasar, and then compare that value to the red-shift-based distance, we might resolve this puzzle.

4.

It's Right *Over* Your Nose

ALL RIGHT, so we believe that there could be older, smarter races out among the stars. Maybe they've even visited here. Perhaps they're watching us with some gentle amusement as we sweat over our dinky little Apollo missions and Mars probes.

When are we going to get some evidence about these alien races—some cold, hard facts that show they really exist?

Many astronomers and cosmologists will give you statistics. They'll state that out of the billions and billions of stars in the universe, even if intelligent races arose on only one out of every hundred billion, there would still be a huge number of intelligent aliens out there. But we're not interested in statistics and speculation now. We're after *evidence*—something we see, hear, taste, touch, or smell.

And we want the evidence now, for us, not our descendants.

We talked about Project Ozma in Chapter 2. That

was an attempt to find such evidence. But Ozma was pitifully understrength to do the job. Our present-day radio equipment can barely detect intelligent radio broadcasts—if any exist—from the nearest fifty or so stars. Make it five hundred or five thousand if you want to. There are a hundred billion stars in the Milky Way galaxy. Unless intelligence is *very* commonplace, the chances of getting to chat with alien creatures on the radio during our own lifetimes are something like fifty in a billion, at best.

All right. Somebody's out there—we hope. But probably not close enough to reach by interstellar phone. So we run smack into the starflight problem again. If we have any hope of seeing or hearing them, either they have to get close enough to make at least a radio contact or we have to go out and find them.

Maybe they are out there, flitting around among the stars, but we just don't know it. Maybe we've actually seen their starships without realizing it. What would a starship look like, from Earth?

Let's try to construct a starship mentally and see if we can find anything in the heavens that fits the description. After all, we have some fairly decent telescopes and radio receivers. Maybe, if we know what to look for, we can come up with a hunk of evidence that shows they're really out there.

We must assume that interstellar ships will be propelled by some form of rockets. We're forced into this. No other propulsion system that we know of today can move a vehicle through space. Except solar sailing, in which you allow the minuscule pressure of starlight to push you along. But solar sailing is incredibly slow. It would take a ship hundreds of years to get from here to Pluto. Count it out as an interstellar propulsion system.

Perhaps a starship would have some form of propulsion that we don't know about—antigravity, or something equally far out. But if we don't know how it works, we don't know what to look for. There could be a sky full of such ships and we'd never realize it.

So we'll have to live with rockets.

Dr. Edward Purcell, a Nobel laureate in physics from Harvard University, tackled the very same problem a few years ago. He worked out the mathematical foundations for interstellar flight; it is published in a book called *Interstellar Communications* (1963). But Dr. Purcell did the job in order to show that interstellar flight is not only impossible, it's hogwash—pure and unadulterated!

He first pointed out that the best you could hope for was a speed of about 99 percent of the speed of light. Fine, we can accept that. As we saw in Chapter 1, relativity theory shows that you can't go faster than light, but at speeds close to light speed there's a time-stretching effect that allows you to cover enormous distances while hardly aging a moment. Combine that with cryogenically suspended animation during the dull portions of the trip, and you've got the possibility of exploring practically the whole known universe within a human lifetime.

But how do you get to that speed? Purcell showed that if you use nuclear fusion engines—even fusion engines that are 100 percent efficient—the rocket ship needs about 1.6 billion tons of propellant for every ton of payload. Billion. A bit uneconomical.

So Purcell looked into the possibilities of using an antimatter drive for the rocket.

Antimatter was first predicted theoretically, and then discovered in experiments involving huge nuclear accelerators—"atom smashers" such as cyclotrons and synchrotrons. Whereas a normal electron has a negative electrical charge, an antielectron has a positive charge and is called a *positron.* A normal proton carries a positive charge, an antiproton is negative. For every normal type of subatomic particle there is an antiparticle.

Antimatter has the interesting property of reacting violently when it contacts normal matter. Both the normal matter and antimatter are completely annihilated and transformed into energy.

In contrast, our hydrogen fusion reaction turns only 0.7 percent of the original hydrogen's matter into energy. A matter-antimatter collision turns 100 percent of the material into energy.

So Purcell examined the possibilities of using matter-antimatter reactions to drive a starship. He found that you need "only" 40,000 tons of propellant—half of it antimatter—for every ton of payload.

But two other problems arise. First: how do you hold antimatter? It can't touch any normal matter, or *boom!*

Perhaps a strong magnetic field—a "magnetic bottle"—could do it. Second, the rocket exhaust of an antimatter drive would pour out some 10^{18} watts of gamma rays. That's a billion billion watts of gamma radiation. This is more energy than the Sun lavishes on our Earth—and sunlight is far more gentle than gamma radiation. If you turned on that kind of engine, you'd bake Earth—or whatever planet you're close to—to a fine dead ash.

Purcell concludes, "Well, this is preposterous. . . . And remember, our conclusions were forced on us by the elementary laws of mechanics."

Preposterous? That's his opinion. It would have been Leif Ericson's opinion if one of his Viking cohorts had shown him the blueprints for a nuclear submarine. It would have been Orville Wright's opinion if he had seen sketches of a swing-wing supersonic jet plane.

All that Purcell's equations really show is that starships should be bulky—huge. And as for radiating 10^{18} watts—marvelous! That kind of light bulb should be visible over long distances and help us to find starships, if they're out there. It's probably safe to assume that anyone smart enough to build a starship might also be smart enough to coast away from planetary neighborhoods before lighting up his main engines.

And, of course, the Bussard interstellar ramjet gets around the propellant problem almost entirely.

But there's another consideration that leads to the conclusion that *anyone's* starship is going to be huge—the time problem.

As we saw in Chapter 1, all starflights are going to be one-way trips, in a sense. Thanks to the time-dilation effect at near the speed of light, you can cover thousands of light-years in the subjective twinkling of an eye, but when you return to your home world, thousands of years will have elapsed there. Even in a very, *very* stable society, things would have changed so much that you'd be out of place. And even if your friends have tremendous life-spans, either they would be so different from you when you return as to be virtual strangers or they would be the biggest bores in the galaxy. People change, and cultures change, over the millennia.

So interstellar voyages are going to be one-way voyages, in effect—unless our concept of the universe is glaringly wrong.

This means that a starship will become all the home that its crew ever knows. Which, in turn, means that the crew's family is going to be aboard. The ships will be little cities of their own—and maybe not so little, either. For just as the Old Testament patriarchs begat new generations, interstellar families are going to grow.

Several thinkers have mentioned in the past that a hollowed-out asteroid might make a good spaceship. Why not consider a larger body, something the size of the Moon or Mars? There would be plenty of room for families and cargo, and lots of hydrogen fuel locked away in the planet's bulk. All the natural resources of a full-sized world would be right there. Sure, the planet-ship would be getting smaller all the time, but you could probably pick up other unpopulated chunks in your travels. In fact, the moons of ice-giant planets such as Jupiter might well be perfect fuel tanks for interstellar ships—little more than fat balls of hydrogen ice.

The starship crew would have to live underground when they're in between stars, but they'd have to live indoors in a factory-built ship anyway. At least, on a reasonable-sized planet, when they got close enough to a warm star they could come outdoors just as soon as their atmosphere thawed out.

The propulsion system that pushes a moderate-sized planet through interstellar space at relativistic speeds (close to light speed) would have to be so powerful that it boggles the imagination. As we've already seen, it staggered at least one Nobel Prize winner. But it's not beyond the known laws of physics! Certainly, we can't build such a rocket engine now; but there's no fundamental law of physics that says it's impossible to build such an engine.

All right, now we know what to look for. At least, we think we know one of the things that we might want to look for. Is there anything resembling a planet-sized starship, using fusion or antimatter rockets, within sight of our telescopes?

Well, what would it look like through a telescope?

Most likely, what we'd see would not be the ship itself, but its exhaust plume, a huge, hot glob of ionized gas, which physicists call a *plasma*. The plasma would expand from the ship's rocket nozzles to enormous dimensions in the hard vacuum of interstellar space. The plasma would be moving at speeds close to that of light, and so would show huge red shifts. And, unlike any natural heavenly body, the plasma exhaust might fluctuate unpredictably as the ship changed course or speed.

Over the past dozen years, the entire astronomical community has gone out of its head trying to figure out what the "quasi-stellar objects," or quasars, might be.

Quasars show enormous red shifts, amounting to speeds of close to 90 percent of the speed of light. Because of these red shifts, astronomers at first thought that the quasars were out at the farthest edges of the observable universe, and their red shifts are caused by the general expansion of the universe.

But quasars twinkle! Some of them brighten and dim over the course of a year or two, others in several weeks or days. A few have been seen to change brightness within a few minutes.

Partly because of this twinkling, many astronomers have leaned toward the idea that the quasars are relatively close by, perhaps not far from the Milky Way galaxy, perhaps actually within it. However, most of the evidence available points to the conclusion that the quasars are at least some distance outside the Milky Way, probably on the order of a hundred million light-years distant. This is still "local," compared to the "cosmological" distances of billions of light-years that were originally assigned to them.

The quasars are apparently composed of very hot gases, plasmas, that are strongly ionized at temperatures of some 30,000 degrees Kelvin. The actual size of the quasars is not yet known. If they're cosmologically distant, then they must be close to the sizes of galaxies. But if they're close to the Milky Way or even inside it, they could be as small as star clusters or even smaller.

Neither cosmologists, astronomers, nor physicists have been able to explain what produces the titanic power output of the quasars. Their light and radio-wave emissions are beyond anything that known natural physical processes can explain. Ordinary physical processes, such as the hydrogen fusion reactions that power the stars, just won't fill the bill. Something else must be burning inside the quasars. A few scientists have suggested matter-antimatter reactions.

Could the quasars be powered by fusion reactors of the type that we would build someday to drive starships? They would run much hotter than the fusion reactions that power the stars. Or might the quasars truly be driven by antimatter reactions?

But if the quasars are starships, and what we're seeing is part of the normal interstellar traffic of the Milky Way galaxy, how come all we see are *red-shifted* quasars? A red shift means the object is moving away from us. Why don't we see any blue-shifted quasars, that is, starships heading toward us?

Maybe we don't see blue shifts because we're out toward the galaxy's edge, and most of the starship traffic is in the star-rich central regions. More likely, though, the answer is that such blue shifts would be very difficult to detect on Earth.

The plasma of the quasars, whether they are the exhausts of starship rockets or not, are inherently very hot, and very blue in color—ranging into the ultraviolet. Most UV wavelengths don't get through our atmosphere—the ozone layers up high in our atmosphere filter out almost all ultraviolet. Only a little UV gets through, and that's what suntans us.

The reason we can see any quasars at all is that their enormous red shifts move the ultraviolet radiation down into the wavelengths of visible light, which do penetrate our atmosphere quite nicely.

Now, a blue-shifted quasar would have its light shifted the other way—from blue and ultraviolet into the far UV, X-ray, and gamma ray wavelengths. None of these wavelengths gets through our atmosphere. So blue-shifted quasars would be quite invisible to us—from the ground. Special ultraviolet detectors placed aboard some of our orbiting astronomical satellites, however, have picked up many, many UV objects that are entirely new to the ground-dwelling astronomers. Could they be blue-shifted quasars? Starships heading our way?

And if they are, should we be doing something to attract their attention?

Of all the improbabilities discussed in this book, the idea that quasars might be starships is by far the least likely to be true. But it's a good mental exercise, both to pose the possibility in the first place, and then to try to poke holes in it. Actually, no one has yet come up with a satisfactory explanation for just what the quasars might be, natural or otherwise. But we'll look at them again in Chapter 6.

5.

Galactic Geopolitics

SO FAR, we have examined the possibilities of building starships, the question of why we haven't been visited by intelligent extraterrestrial creatures, and the evidence that we may indeed be watching interstellar commerce through our telescopes.

Now let's assume that contact with another intelligent race is inevitable. Sooner or later they will come to visit us, or we'll stumble into them once we get our starships cruising across the Milky Way.

It seems almost certain, incidentally, that we won't find another intelligent species among the planets of our own solar system. Mars and Venus have been blasted from our hopes by the pitiless advance of knowledge, thanks to space probes. Mercury, Pluto, and our own Moon were never really counted on as habitats for intelligent races. And the Jovian planets—Jupiter, Saturn, Uranus, and Neptune—are *too* alien for us. More on them later.

If we find another intelligent race, it will be out

among the stars. Assuming that brainy aliens are out there, what are the chances of having any meaningful, fruitful contacts with them? Not just radio chats, not just an occasional awe-inspiring visit. Real, long-term, continuous interaction, the way the United States interacts with the other nations of Earth—trade, cultural interpenetration, tourism, politics, war.

This all depends, of course, on attaining starflight. More than that, it has to be fast, cheap interstellar transportation. Otherwise there can be no large-scale interactions, no politics or trade, between us and them.

Look at a parallel from Earth's history.

Since at least Roman times, western Europeans knew that China and the Orient existed. In the Middle Ages, Marco Polo got there and back, spreading wondrous tales that grew each year. But Europe didn't interact with China in any significant way. True, Europe engaged in trade with the Arab Middle East and obtained goods from China through Arab middlemen. The Middle East was close enough for Europeans to reach on foot if they had no other way to get there. Europe traded with the Middle East, exchanged scholarly works—which is why most of the stars in the sky have Arabic names—and engaged in the pious slaughters called the Crusades.

But there was no direct trade, and no conflict, with China. Once deep-ocean sailing vessels were perfected, though, Europe did indeed contact China directly and treated the Orient to Western technology, trade, disease, and war. Today, of course, with intercontinental rockets and instant communications, everybody on the globe can interact politically with everybody else.

The same rules will apply to interstellar politics. There may be glorious civilizations in the Orion complex, or even as close as Alpha Centauri. But we know less about them than Hannibal knew about China. No action.

Yet even today it is possible to visualize starships based on technology that is tantalizingly close to our grasp. If and when we can make trips to the nearest stars within a human lifetime, we'll have reached the Marco Polo stage of interstellar contact—adventure, strange tales, and stranger artifacts. But no lasting political relations, for better or worse, with the neighbors.

There would be little tourism, except of a scientific variety, when a person could visit the exotic land only once in a lifetime, and the trip would consume a fair portion of his life-span. It is also hard to picture commerce and trade relations based on one ship per human generation. That's more like a cultural exchange. And even the sternest, most fearless and ruthless general might feel a bit foolish about mounting an attack when he knew he could never see the outcome in his own lifetime.

But the real importance of Marco Polo's adventure was the spur it gave to Prince Henry the Navigator and others, including Christopher Columbus. And the importance of the first interstellar contact will be the stimulus it gives to us on Earth.

Now, if you corner a theoretical physicist, the chances are that you can start him mumbling about tachyons and things that go faster than light. Einstein's light barrier is starting to look—well, not leaky, perhaps, but at least a little translucent. Perhaps one day ships will be able to zip among the stars at speeds far greater than light.

Since we're dealing with improbabilities in this book, let's consider this one. With faster-than-light ships, we can get just as close and chummy with our stellar neighbors as we are today with the Chinese.

But we must realize that there will be many races out among the stars that we simply *cannot* interact with in any useful way, even though we may be able to reach them physically.

Remember the "postage stamp" analogy from Chapter 2? We may, for example, find races much younger than our own, with a correspondingly simple technology and social development. Aside from letting them worship us as gods, there's probably little that we could do for them—or they could do for us.

Certainly we would want to study them and learn more about how intelligence and societies evolve. That would be best done from orbit, where we could remain "invisible" and not disturb them in any way. What could they offer us, except for their own artifacts or bodies? The artifacts might be interesting as examples of alien art. And no matter how lopsided or gruesome they appear, there will arise at least one art critic who will explain the hidden aesthetic values that everybody else has missed and sell the stuff at a huge markup.

And their bodies?

We wouldn't use them for meat, for a number of reasons. If their bodies contained some precious chemical substances that couldn't be found elsewhere—the key to immortality or something equally exotic, we would be in a lovely ethical bind. But the chances for that sort of situation are vanishingly small. We certainly would not need muscular slaves in our technological society—electricity is cheaper. And we have laws about such things, anyway.

And what could we offer our younger neighbors? Only the things that would destroy their culture as surely as western Europe destroyed the American Indians. We hope that by the time we reach such a race, we'll have learned not to interfere with them.

If we should try to meddle with a race that's only slightly younger or technologically weaker than we are, their reaction could very well be the same as the Indians'—they would resist us as strongly as they could, probably with guerrilla warfare. We're finding out in Vietnam exactly what Custer learned nearly a century ago—that "unsophisticated" and "simple" people can use our own technology very effectively against us. But the Indians were either killed or absorbed into our culture, and the Vietnamese are going through the same process. That part of the world will never again be a simple, unspoiled, isolated Asian backwater. The same thing would probably happen to a younger race that fights against us: the very act of resistance will destroy their native culture.

What happens when we contact a race much more advanced than we are? The same situation, only in reverse. We would have precious little to offer them, except possibly curiosity value. And they would be wise enough not to tamper with us. We hope. Playing cowboys and Indians is no fun when you're on the foredoomed side.

A really far advanced race would most likely go its own way aloof and serene, even if we tried our hardest to make friends. The picture that comes to mind is a puppydog chasing a monorail train.

That leaves us with races that are more or less at our own stage of development, intellectually, morally, and technologically. *That's* where the fun—and the danger—will be.

How much of a range is covered by "more or less" is rather hard to say. For a thumbnail definition, let's put it this way: We will interact strongly with races that have something to gain from us, and vice versa. Cavemen and angels have so little in common with us that they won't affect us very much, nor will we affect them. But other humans, even if they're purple and have sixteen legs, will provide the interstellar action.

Further, the races we interact with will probably come from planets enough like our own to make this Earth attractive to them. And their home worlds will

Hale Observatories

A lovely spiral galaxy, M 81, in the constellation of Ursa Major (the Great Bear). Our own Milky Way galaxy probably looks something like this, with a core of densely packed reddish stars, and spiral arms of younger, bluish stars.

similarly be attractive—or at least bearable—to us. This is why, even if intelligent Jovians exist under Jupiter's clouds—or Saturn's or Neptune's or Uranus', we probably will interact with them about as much as we do with the denizens of the Marianas Trench. There's just no common meeting ground. We don't have political relations with dolphins, even if they are as intelligent as we are. We have nothing to trade or fight over.

So it boils down to this: Although we may meet many strange and marvelous races among the stars, if they are physically or intellectually far removed from us we will have little but the most cursory of contacts with them—except for scientific expeditions.

Yet the races that can stand on our planet in their shirt sleeves, or at least a minimum of protective equipment, and have a technology of a roughly similar level to our own, will be the races that we will talk with, laugh with, trade with, and fight with. It may be that intelligent life is too thinly scattered through the Milky Way's stars for us to expect to find such a race close enough to us—close enough in distance and maturity—to make interstellar politics likely.

Going back to the "postage stamp" analogy again, just what are the chances of meeting another intelligent race that is at our own stage of development, within the few centuries represented by the postage stamp's thickness? Or even within the 10,000-year thickness of the dime? Below that dime are cavemen. Much above the postage stamp, and we're in the realm of highly advanced civilizations that would regard *us* as cavemen.

The chances for meeting neighbors that we can truly interact with seem mighty slim, when you think about that postage stamp. But let's look around anyway and see what the real universe holds for us.

There are 37 stars within five parsecs of the Sun. Of these 37 stars, 27 are single, 8 are binary, or double, stars, and 2 are triples. Four of these stars are known to have "dark companions"—bodies of planetary mass that are too faint and small to be seen. In fact, two of the nearest five stars have planets. Since planets are extremely difficult to detect, we might suspect that there are plenty of them orbiting the farther stars, but we just cannot perceive them from here.

If the population density of stars is about the same as we go farther away from the Sun, then there should be something like 300,000 stars within 100 parsecs of us, and some 300 million stars within 1,000 parsecs. As we have seen, the Milky Way galaxy as a whole contains about 100 billion stars. Our galaxy is roughly 30,000 parsecs in diameter, and our solar system is some 10,000 parsecs out from the center.

We have no way of knowing how rare intelligence is. But in every cosmological test that has been applied to the Earth and solar system so far, we find absolutely no evidence for our own uniqueness. Quite the opposite. The Sun is a rather average star. It appears that planets form around stars naturally. Planets at our temperature range from their star should turn out to look roughly like Earth, with plenty of liquid water. Life on those planets should be based on carbon, oxygen, and water, making use of some of the most abundant materials and most energetic chemical reactions available. Given enough time, the natural forces that led to the evolution of life on Earth would lead to similar results on similar worlds.

The real question is: What are the ages of the stars around us? If they are about the same age as the Sun, we might expect to find interesting neighbors.

The Sun's age has been pegged at roughly 5 billion years. This is based chiefly on estimates of the amount of the Sun's original hydrogen that has been converted into helium through the hydrogen fusion processes that make the Sun shine. In turn, these estimates are based largely on theory, since no one can look inside the Sun and actually measure the ratio of hydrogen to helium there. In fact, no one knows how much helium, if any, the Sun had when it first began to shine. But 5 billion years is a reasonable guesstimate, and it tallies well with the ages of the oldest rocks of the Earth, the Moon, and the meteorites.

Many of the stars around the Sun are clearly much younger. Table 1 shows the classes of typical stars according to their spectrums, together with estimates of their stable life-spans. By "stable life-span" we mean the length of time that the star is on the Main Sequence.

To explain: Stars go through an evolutionary path, a life-span, much as do living creatures. In the vastness of the Milky Way, stars are constantly being born and dying. The evolutionary path for an ordinary star, such as the Sun, goes like this:

1. A "protostar" condenses out of interstellar gas and dust. The protostar, a dark clump of mostly hydrogen, is about a light-year wide. It contracts rapidly, falling inward on itself under the gravitational force of its own mass. As it contracts, it naturally gets denser and hotter. Its interior temperature rises sharply.

TABLE 1. *Spectral Classes and Life-spans of Stars*

SPECTRAL CLASS	SURFACE TEMPERATURE (°'S KELVIN)	COLOR	STABLE LIFE-SPAN	EXAMPLE
B	11,000–25,000	blue	8 to 400 million yrs.	Rigel, Spica
A	7,500–11,000	blue-white	400 million to 4 billion yrs.	Sirius, Vega
F	6,000–7,500	white	4 to 10 billion yrs.	Canopus, Procyon
G	5,000–6,000	yellow	10 to 30 billion yrs.	Sun, Capella, Alpha Centauri A
K	3,500–5,000	orange	30 to 70 billion yrs.	Arcturus, Aldebaran, Alpha Centauri B
M*	below 3,500	red	more than 70 billion yrs.	Alpha Centauri C, Barnard's star

* Red supergiant stars such as Betelgeuse and Antares are not Main Sequence stars, therefore their stable life-spans in no way correspond with those of the red M-class dwarfs shown on this table.

2. When the density and temperature at the core of the protostar reach a critical value, hydrogen fusion reactions are triggered. The gravitational collapse stops, because now heat and light produced by fusion are making outward-pushing pressures that balance the inward-pulling gravity. The star shines with fusion energy, it becomes a stable member of the great family of stars that astronomers call the Main Sequence. Its size and surface temperature will remain stable as long as hydrogen fusion provides the star's energy source.

3. The bigger and more massive the star is to start with, the hotter it is, and the faster it runs through its hydrogen fuel supply. When the hydrogen runs low, the star begins burning the helium "ash" that is left in its core. Helium fusion, producing oxygen, neon, and carbon, runs hotter than hydrogen fusion. The star's central temperature soars, and the outer layers of the star are forced to expand. The star is no longer a Main Sequence member—astronomers call it a red giant. Soon, in astronomical time scales, the helium runs low, and the star begins burning the heavier elements in its core. The star continues to create, and then burn, constantly heavier elements. All the while, the core is getting hotter and the star's outer envelope is swelling enormously. When the Sun goes into its red giant phase, it may get so large that it swallows its inner planets—including Earth.

4. Eventually the star reaches a critical point. It ex-

plodes. There are several different types of stellar explosions, and several courses that the evolutionary track might take from there. More on these possibilities in Chapter 6. For now, we need only realize that the eventual outcome of this stellar violence is a white dwarf star (a fading dim star about the diameter of the Earth or smaller) or an even tinier, denser neutron star. Neutron stars are probably no more than 10 kilometers across, yet they contain as much material as the Sun! They are fantastically dense. The pulsars, whose uncannily precise pulses of radio energy led briefly to the "LGM theory," are probably fast-spinning neutron stars.

As we can see, a star remains stable for only a certain finite period of time, depending on its mass and temperature. After that, things get pretty dramatic for any planet-dwelling life nearby.

Hot blue giants such as Rigel and Spica won't be stable for more than a few hundred million years. While this is a long time in terms of human life-spans, it is an eyeblink in terms of evolution. This means that such stars cannot be more than a few hundred million years old. In all probability, the dinosaurs never saw Rigel. It wasn't there yet.

We know that it took about 5 billion years for intelligent life to develop on Earth. As a rule of thumb, lacking any better evidence, we can say that we shouldn't expect to find intelligent life on planets circling stars that are less than 5 billion years old. So Rigel and the other young blue giants can probably be ruled out as possible abodes for intelligent life.

The stars that are smaller and cooler than the Sun, such as the K and M dwarfs, have much longer life expectancies. But are they older than the Sun? There's no easy way to tell.

We might be able to get some clues to their ages by looking farther afield. Consider the "geography" of the Milky Way.

The Milky Way is, of course, a spiral galaxy very much like the beautiful nebula in Andromeda. The core of our galaxy is presumably thick with stars, but we never see the core because it is hidden behind thick clouds of interstellar dust. Radio and infrared observations have been able to penetrate the clouds to some extent, and observations of the central regions of other galaxies show that they are so rich with stars that these stars are probably no more than a single light-year from each other, at most.

Stars in the core of a galaxy are also presumably much older than the Sun—red giant stars are common there, and astrophysical theory shows that stars become red giants only after they have used up most of their hydrogen fuel and have ended their stable Main Sequence phases. Also, in the cores of galaxies there are no young, hot, blue giants such as Rigel and Spica. These are found only in the spiral arms of galaxies.

Because the core regions of spiral galaxies seem to have different types of stars, predominantly, than the spiral arms, astronomers refer to the two different stellar constituencies as Population I and Population II. And thereby they sometimes cause confusion.

Population I stars are the kind our Sun lives among. These are the youngish stars of the spiral arms. Their brightest members are the blue giants. Population I stars contain a relatively high proportion of elements heavier than hydrogen and helium. Although the proportion of heavy elements hardly ever amounts to more than one percent, the Population I stars are said to be "metal-rich."

Population II stars are those found in the core regions of a galaxy. They are old, their brightest members are red giants, and they are mostly "metal-poor."

The heavy-element content of a star is an important clue to its history. Why are the stars in a galaxy's core metal-poor and the stars in the spiral arms metal-rich? Because the elements heavier than hydrogen have been created inside the stars. It works this way:

Consider the Milky Way before there were any stars. Cosmologists have estimated that the Milky Way is between 10 and 20 billion years old, that is, some two to four times older than the Sun. Presumably, the whole universe is the same age as the Milky Way. But when you are dealing with tens of billions of years, the numbers tend to get imprecise and hazy.

Regardless of the exact age of our galaxy, it began as

an immense dark cloud of gas at least 30,000 parsecs across. The gas might have been entirely hydrogen, or it might have been a hydrogen-helium mixture. Where this gas originally came from is a mystery that cosmologists argue about, but no one has been able to prove which side of their argument is right—if either.

The first stars to form had no elements heavier than helium in them. Perhaps nothing more than hydrogen. All the heavier elements, from lithium to iron, were "cooked" inside these stars as they went from hydrogen-burning to helium-burning to heavier-element-burning. Some of these stars exploded, in the last stages of their lives, with the titanic fury of the supernova. In those star-shattering explosions, still heavier elements were created, beyond iron, all the way up to uranium and even beyond that. There's some evidence that the so-called "man-made" element, Californium 254, was present in the supernova of A.D. 1054, which we know today as the Crab nebula.

So the first generation of stars in the Milky Way began with only hydrogen—perhaps laced with a smattering of helium—and eventually produced all the heavier elements. And the stars threw these heavier elements back into space, where they served as the building material for the next generation of stars. The explosions that marked the death throes of the first-generation stars enriched the interstellar clouds with heavy elements. It is from these clouds that new stars are born.

Judging by the heavy-element content in the stars, most astrophysicists estimate that the Sun must be a third-generation star, a grandson of the original stars of the Milky Way. The elements inside the Sun today were once inside other stars. The atoms that make up the solar system were created inside other stars. The atoms of your own body were made in stars. We are truly star children.

Beware of a clash of jargon when we talk about generations of stars and Population I or II. Population I stars are the younger, late-generation stars. Population II stars are the older, early-generation stars. II came before I, historically.

What has all this to do with meeting the neighbors?

Just this: The first-generation stars *could not produce life.* At least, nothing that we would recognize as life. There was no carbon, no oxygen, no nitrogen . . . nothing but hydrogen and perhaps some helium. If those first stars had planets, they would all be frozen ice balls of hydrogen, somewhat like Jupiter but not so colorful, because there would be no ammonia or methane or any other chemical compounds to cause gaudy streaks of colored clouds such as we see on Jupiter and Saturn. There wouldn't even be any water. Not yet.

Second-generation stars? It's possible that they would have most of the heavier elements, including the carbon, oxygen, nitrogen, potassium, iron, and such that we need to develop for life. Planets of such stars might be able to support life, even our own kind of life, if these heavier elements were present in sufficient quantities. And if life has appeared on such planets, there's no reason to suppose it wouldn't eventually attain intelligence. Certainly the long-lived red dwarf stars provide plenty of time for intelligence to develop—5 billion years plus.

Let's grant that an intelligent race could arise on the planetary system of a second-generation star. Could such a race develop a high civilization and technology? It all depends on the abundance of natural resources. Fossil fuels such as coal and oil should be plentiful, since they are the result of the biodegrading of plant and animal remains. But what about metals? Our technology here on Earth is built around metals. Even our history rings with the sounds of the Bronze Age, the Iron Age, the Steel Age, the Uranium Age.

Astronomical evidence is indistinct here. Theory shows that second-generation stars should have a lesser abundance of metals than we third-generation types have. But certainly there should be some metals on second-generation planets.

How much metal is enough? There's no way for us to tell. Planets of second-generation stars might have iron mountains and gold nuggets lying on the open ground. Or they might have very little available metal. Our own Jupiter might easily have more iron in it than Earth does. But if it's there, the iron is mixed with 317.4 Earth masses of hydrogen, helium, methane, ammonia, and whatnot. Try to find it! And get at it!

If there are planets of second-generation stars where heavy metals—iron, copper, silver, tin, gold—are abundant and available, those planets could be sites for highly advanced civilizations. But suppose intelligent races arise on planets where heavy metals are not available? What then?

First, we should clearly realize that intelligence per se does not depend on heavy metals. *Life,* though, does, to some extent. There's an atom of iron at the core of every hemoglobin molecule in your body. And hemoglobin is what makes your red blood cells work. So without iron, and certain other heavy metals, we wouldn't be here!

Mankind rose to intelligence before he discovered heavy metals. He used wood, clay, rock, and animal bones for his first technology. In a way, man went through a Ceramics Age, working mostly with clay, before he found metals. In fact, it was wood and ceramics that allowed man to handle fire safely and usefully. Only after fire had been tamed could men start to use metals on a large scale.

The history of man shows that once metals became available, we took a giant leap forward. Metals allowed men to build effective plows. And swords. And chariots. Even today our skyscrapers and computers and engines and spacecraft and weapons and household appliances are made mainly from metals. Metals are strong, tough, and cheap. They are rather easily found and easily worked, even with low-grade fire.

Hale Observatories

A globular cluster, containing hundreds of thousands of stars, in the constellation Canes Venatici (the Hunting Dogs). There are more than a hundred such globular clusters forming a "halo" around the Milky Way. They contain very aged, metal-poor red giant stars.

Could a race build skyscrapers and spacecraft without metals? Well, today there are many "space age" materials such as plastics and boron-fiber composites. But the machinery that produces them is made of stainless steel, copper, brass, etc. Modern technology is showing that there are nonmetallic materials that can outperform metals in strength, weight, and many other performance parameters. But these materials couldn't have been developed before an extensive Metal Age technology. Cavemen, or even the ancient Greeks, could not have produced boron-fiber composites or modern plastics. They didn't have the metals to produce them with!

Would a metal-poor second-generation intelligent race be stymied in its attempts at technology? Who can say? All we know for sure is that *our* technology certainly depends on metals, and until metals were available, our ancestors had no civilization or technology higher than Neolithic.

Another vital point. While we have nothing but the history of our own race to go on, it looks very much as if the whole world of electromagnetic forces would never have been discovered without metals such as iron and copper. Man's discovery of magnetism depended on the abundance of iron on this planet. And from the very beginnings of our experiments with electricity, we used lead, zinc, copper, brass, etc. It's hard to see how the entire chain of study and use of electromagnetic forces—from Volta and Faraday and Hertz through to radio telescopes and television and superconducting magnets—could have happened on a metal-poor planet. And where would our technology be without electricity? Back in the early nineteenth century, at best.

So what about the metal-poor second-generation races? It just might be possible to build a complex technology completely out of nonmetals. But tribes on Earth that never had easy access to heavy metals have never developed a high technology. Coincidence? Maybe.

Could a strong technology be built around the lighter metals, such as lithium, beryllium, or boron? Ironically (pardon the pun) those metals are much less abundant in the universe than the heavier metals (iron and up). And for good reason. The light metals make excellent "fuel" for the nuclear fusion reactions inside stars. They are used up inside a star before it explodes and spews out its material for later generations. So the chances of having a sophisticated civilization based on light metals seem slim indeed.

If our own history is any guide, it is the heavy metals that lead to high technology. And they also form a natural gateway into the world of electromagnetic forces and the whole concept of "invisible" forces that act over a distance: magnetism, electricity, gravity, nuclear forces. We can trace a direct line from man's use of heavy metals to electromagnetics, nuclear power, and, we hope, beyond.

For second-generation stars the situation is much

cloudier. Either they have enough heavy metals to develop a high technology or they don't. If they do, their races are much older and presumably wiser than we are. Which means they probably won't interact with us at all. We would probably bore them to tears, or whatever they have in place of tears.

Second-generation races that don't have metals are no doubt gamboling innocently through some local version of Eden, and we should leave them strictly alone.

There goes the long-standing science-fiction vision of an immense galactic empire, run by the older and wiser races of the Milky Way's ancient core regions. Like the "steaming jungles" of Venus and desert "cities" of Mars, the empire at the center of the galaxy simply doesn't exist. The first-generation stars produced no life. If there are second-generation intelligences around, chances are they're either *so* far advanced beyond us that empires are meaningless trivialities to them, or they're so metal-starved that they never got past the "Me Tarzan" stage of development.

It's a shame. It would have been pleasant to talk to them—those incredibly ancient, benign, and understanding superbeings from the galaxy's core. It's sort of shattering to realize that, if anyone like them does exist, they wouldn't want to be bothered with our chatterings any more than a crotchety grandfather wants to try to converse with a squalling baby.

On the other hand, science-fiction stories abound in which a race only slightly advanced over us—say, a few centuries—does a very ruthless job of conquering Earth. So maybe we should be glad if there's no one older who is interested in us.

Of course, an older race might be benign. If so, it would probably not reveal its presence to us, for fear of damaging irreparably our culture and our spirit. They would prefer to wait until we could meet them on a more equal footing. The "equality" point might be when we've achieved successful starflight for ourselves.

If an older race is not benign, but aggressive, then it might want to gobble us up before we had reached the stage of starflight. That way, we would be alone and defenseless against them.

So if we should be visited by aliens from another solar system *before* we achieve starflight, my hunch is that their intentions will be far from pleasant—no matter what they say.

But the chances of meeting another race that is even within a few centuries of our present stage of development seem rather remote. And remember, the Sun is one of the oldest third-generation stars around this part of the galaxy. There might not be any older races within thousands of parsecs of us.

Could it be that *we* are the oldest, wisest, farthest-advanced race in this neck of the stellar woods?

Now, that's a truly sobering thought!

6.
When the Sky Falls

WE ARE COMING to one of the strangest twists of all. The subject is still starflight, the way we might reasonably expect to reach out into the vastness of the Milky Way. But we're going to approach our subject from a very different direction. We're going to look at the way stars die—how the Sun and other stars eventually collapse and go dark. For in the death of a star there just might be a path to interstellar travel.

In the past few years, discoveries such as the quasars and pulsars have forced astronomers and physicists to look carefully into the phenomenon of *gravitational collapse:* that is, what happens when a star or even an entire galaxy falls inward on itself, sucked down into a maelstrom of its own making, because of its own gravitational energy.

It turns out that this weird and wonderful domain of gravitational collapse might contain the secret energy source of the quasars. This is a domain where hugely massive stars can wink out and entirely leave this universe, and, just maybe, a domain where we might find the key to faster-than-light travel.

Like so much of the past decade's new astronomical thinking, the discovery of the quasars prompted an intense look at the mechanics of gravitational collapse. The energy output of the quasars is so huge that old ideas about energy production in stars and galaxies had to make way for new concepts. As Table 2 shows, a typical quasar is emitting more visible light energy than a thousand Milky Way galaxies—and as much radio energy as the strongest radio sources known!

TABLE 2. *Optical and Radio Output of Galaxies and Quasars*

SOURCE	OPTICAL OUTPUT (KILOWATTS)	RADIO OUTPUT (KILOWATTS)
Milky Way (spiral galaxy)	10^{33}	10^{28}
M 87 (elliptical galaxy)	10^{34}	10^{32}
Cygnus A (strong radio source)	10^{33}	10^{35}
Typical quasar	10^{36}	10^{35}

Just to put those very large numbers in some sort of context: M 87 is one of the largest and (optically) brightest galaxies. It probably contains a trillion (10^{12}) stars. Yet the typical quasar is a hundred times brighter. Cygnus A is one of the strongest radio sources in the sky, and the typical quasar is just as powerful in radio output. The energy output for quasars shown in Table 2 is roughly equal to the energy emitted in *10 billion* supernova explosions, or the energy obtained by the matter-antimatter *total annihilation* of 10 million stars.

Could there be 10 billion supernovas blazing in chain reaction in a quasar? Or 10^7 solar masses of matter and antimatter merrily destroying each other in a million-year-long celestial fireworks display?

Although the quasars seem to be emitting as much energy as (or more than) the most powerful optical and radio sources in the heavens, they are apparently much smaller than any galaxy. How can the energy of a thousand galaxies be packed into a space that's considerably smaller than a single galaxy?

So far, we have assumed that the quasars are "cosmologically" distant—1 billion light-years or more away from us. This puts them out at the edges of the observable universe, for the most part. Astronomers made exactly the same assumption in the early 1960's, when they first realized that the quasars were very different from anything they had previously seen.

The reason that the quasars were assumed to be "cosmologically" distant is that they show tremendous red shifts. The farthest known true galaxy, 3C 296, has a red shift that is estimated to represent a speed of recession of 36 percent of the speed of light. This works out to a very rough distance estimate of 5 billion light-years. Most quasars do much better. A handful of quasars have such huge red shifts that they're apparently moving at about 90 percent of light speed; this yields a distance judgment of more than 10 billion light-years.

Remember, this red-shift method for gauging cosmic distances is at best very rough and depends on an interlinking chain of assumptions: (1) that the observed red shifts are Doppler shifts, caused by the objects' rushing away from us, (2) that the reason they're moving away is that the whole universe is expanding, and expanding *uniformly,* so that the farther away a galaxy or quasar is from us the faster it's receding, and therefore, (3) that the larger the red shift, the faster the object is receding, and thus the greater its distance from us.

Hubble showed that if you make a graph plotting the brightnesses of galaxies against their red shifts, the relationship is a beautiful straight line. This is extremely powerful evidence that the red shifts are truly related to distance. But if you plot the brightnesses of the known quasars against their red shifts, you don't get a straight line at all. You get a wild shotgun pattern, with no apparent relationship to anything except confusion.

This led Fred Hoyle, the British cosmologist, to begin wondering if the quasars' red shifts might be completely unrelated to distance. Maybe the quasars are not cosmologically distant at all, but relatively nearby, perhaps only a few million light-years away, at most.

Hoyle needed "local" quasars if he was to save the well-known steady state theory of cosmology. For the quasars, if cosmologically distant, tended to show that the universe was definitely very different 10 billion years ago than it is today. If you count the quasars and galaxies together, the universe was more densely packed with such objects 10 billion years ago than it is now. All this destroys the steady state theory, which claims that the universe has always been about the same as it is now.

The opposing big bang cosmology pictures the origin of the universe in one cataclysmic burst of energy. Many cosmologists looked on the quasars as evidence for that primal explosion. But if Hoyle could show that the quasars are local objects and not related to events of 10 billion years ago, then the steady state theory might still survive. On the other hand, if it could be shown that the quasars are local and their red shifts not related to distance, then some doubt is cast on the value of red-shift measurements for judging the distances of all the galaxies. Some doubt might even be cast on the very concept of an expanding universe. So the local vs. cosmological quasar argument had—and still has—high stakes attached to it.

So in 1963 Hoyle and William Fowler, astrophysicist from CalTech, proposed that the quasars might be supermassive objects relatively close to our own galaxy. They saw the quasar as being much smaller than a galaxy, perhaps like a globular star cluster, but with a mass 100 million times the Sun's. For lack of a better tag, call it a superstar.

Both the energy output and the red shift of the

Hoyle-Fowler superstar were attributed to gravity. Gravitational collapse provided the basic energy for the superstar's outpouring of light and radio waves, as gravitational energy is converted to electromagnetic energy while the superstar shrinks in size and becomes constantly denser, more compact.

The powerful gravitational field of this supermassive object causes the red shift. As photons work "uphill" against such a strong gravitational field, they are shifted down toward the red end of the spectrum. Similar effects, although much smaller in magnitude, have been observed on the Sun and other stars.

The superstar idea came under immediate attack, as have all theories hoping to explain the quasars. A single object of 10^8 solar masses could not remain stable, said the physicists. All right, said the theory's backers, call it a super-star-cluster, then. It can still be treated as a single object even if it consists of many smaller parts. And, they showed, if the superstar were rotating rapidly enough, it would not break up.

The argument is still going on. Big-bang cosmologists want the quasars to be cosmologically distant. Steady-state people want them local. In all fairness, there have been several other suggestions that the quasars are local. For example, James Terrell of the University of California proposed in 1964 that the quasars might be something like massive star clusters that have been shot out of our own or nearby galaxies. The red shifts, then, would be Doppler shifts caused by the ejected quasars' recession, but would have no relation to cosmology.

That same year, C. R. Lynds of Lick Observatory and Allan Sandage of Mount Wilson and Palomar Observatories showed definitely that the galaxy M 82 was in the throes of an explosion. Its core had blasted itself apart, perhaps as recently as a few hundred thousand years earlier. Could the quasars be "shrapnel" fired out of exploding galaxies?

More on galactic explosions shortly.

The idea of gravitational collapse powering the quasars was not restricted to local-quasar enthusiasts. Even the astronomers and cosmologists who backed the cosmologically distant quasar theory considered gravitational energy as a possible source of the quasars' enormous brightness. In this case, they looked on the quasar as something about the size of a galaxy that is collapsing inward on itself. The energy release could be purely gravitational in origin, or it could come from the collision and explosion of billions of stars in the galaxy's core, as they were squeezed together in the general collapse of the galaxy.

The discovery of the quasars and the earlier realization that the so-called radio galaxies frequently have small regions in their cores from which most of the radio energy emanates led astronomers to begin paying more attention to what's going on in the cores of galaxies.

They dusted off the work done in the 1940's by the

Hale Observatories

The results of a nova explosion. This star, in the constellation Perseus (a legendary Greek hero) exploded in the year 1901. The plasma cloud surrounding the star is expanding into space. The star itself seems none the worse for its experience.

American astronomer C. K. Seyfert, who studied a number of galaxies that have extraordinarily bright cores. Seyfert galaxies, as they are now called, have very active cores in which there is much highly excited loose gas moving with velocities of some 4,500 km./sec. While Seyfert worked exclusively with optical telescopes (radio astronomy was still only a gleam in Grote Reber's eye), more recent radio studies of the Seyfert galaxies show them to be fairly powerful radio sources, with the radio emission coming from those bright, agitated cores. Incidentally, the Seyfert galaxies resemble the quasars in many respects, including the fact that both tend to show sizable variations in light and radio output. But the quasars are at least a thousand times more powerful—if they're cosmologically distant.

By the mid-1960's evidence for galactic explosions began pouring in. Lynds and Sandage showed that M 82 is exploding. Short-exposure photographs of the elliptical galaxy M 87 showed that it has an optically bright spot at its core, with a jet of glowing plasma, some 30,000 light-years long, streaking off to one side! Previous photos of M 87, long-time exposures to catch the faintest star clusters around it, had washed out this feature completely.

There's even a strong chance that our own galaxy suffered a core explosion at least a million years ago. A "halo" of radio-emitting gases observed around the Milky Way could have been ejected from the core in an explosion similar to that of M 82. Some astronomers now believe that most, if not all, of the radio activity in galaxies and quasars is associated with explosions at the core.

What causes galactic explosions? Where does the energy come from? Is it a coincidence that the energy involved in an exploding galaxy, according to most calculations, works out to be very similar to the energy output from the "cosmologically distant" quasars?

As in the case of the quasars, theories for galactic explosions abound. Again, they include outright gravitational collapse, stellar collisions, and/or supernova chain reactions, and matter-antimatter annihilation. Each possibility needs some sort of gravitational collapse to make it work. And again, none of the explanations can answer all the tests and objections that have been brought out.

As you might suspect, all this attention on gravitational collapse as a power source for quasars and exploding galaxies led the astronomers and cosmologists to turn expectantly to the physicists for some answers. And that's just what they got—some answers. Not *the* answers they were looking for, unfortunately, but some fascinating food for further thought.

The physicists had been following the same gravitational-collapse trail from a different starting point. They were studying individual stars in an attempt to explain the evolution of a star—an evolution that sometimes ends in a supernova explosion. It's ironic that the physicists were sniffing along this trail because of their interest in what was, up to the mid-1960's, the most titanic catastrophe known to man: a supernova. And when they bumped noses with the astronomers, it was because the astronomers had found cataclysms 10 billion times mightier.

Let's get away from quasars and galaxies for a while, and start thinking about plain little old stars. Like the Sun. In following the physicists over this portion of the trail, we'll soon enough reemerge into the wider cosmos of pulsars, neutron stars, quasars, expanding and contracting universes, and (as advertised earlier) maybe faster-than-light travel.

Stars begin life with gravitational collapse. The Sun, for example, was a loose cloud of gas and dust some 5 billion years ago. Under its own gravitational forces, the cloud contracted in an astronomical eyeblink (about 50 million years, according to computer calculations) and formed a medium-sized star with some cosmic debris orbiting around it.

Why did that gravitational collapse stop where it did, leaving the Sun with its present almost perfectly spherical diameter of 1.39 million kilometers? Because at the Sun's central temperature of some 20 million degrees Kelvin, hydrogen fusion reactions produce enough gas and radiation pressure to balance the still-present gravitational pull of 2×10^{27} tons of matter.

As we've already seen, in another 5 to 10 billion years, the Sun's hydrogen supply will start to run low. Most of its core will be helium, created from the hydrogen fusion process. So the core will be denser than it is now, and hotter. Its central temperature will rise to some 100 million degrees, and then the helium will begin to fuse into carbon, oxygen, and neon.

At the higher core temperatures associated with this new energy source, gravity must yield somewhat to increased gas and radiation pressure. The Sun's outer layers will expand. The surface of the Sun, the photosphere, will become distended and cooler. The Sun will become a red giant star.

The same routine will be repeated over and again. As the fusion reactions in the Sun's core produce constantly heavier elements, the core temperature rises. The higher the core temperature, the easier to start fusion reactions with the heavier elements, leading to the creation of still-heavier elements, still-higher temperatures, and so on.

Each cycle of new-element-building goes faster than the previous one. Each cycle is bringing the Sun closer to disaster. Through it all, gravity is constantly being outfought by rising gas and radiation pressures, and the Sun's outer envelope becomes hugely distended. Meanwhile, despite the higher core temperatures, the surface temperature still goes down.

This continues until the fusion reactions at the core produce iron. When iron nuclei fuse they produce lighter elements, not heavier ones. The game is over. And gravity, which has been patiently waiting all this time, becomes the victor. The remainder of the star's life will depend more on the always-abiding force of gravity than on any other factor.

What happens then?

Since 1915, when the first white dwarf star was discovered (the "pup" of the "Dog Star," Sirius), astronomers and astrophysicists have assumed that somehow most stars must eventually end as dying white dwarfs. But how does a star go from being a red giant to a white dwarf? (This sounds like a question out of a fairy tale, rather than a problem in nuclear astrophysics.)

Stars have been known to explode. Sometimes they explode rather mildly, in cosmic burps called novas; sometimes dramatically, in supernova explosions that release as much energy in 24 hours as the Sun emits over a billion years. Where do these stellar explosions fit in? Will the Sun explode?

These are the questions that the astrophysicists were working on when the quasar storm struck. One of the leading workers in the field, who has concentrated his studies on the phenomenon of gravitational collapse, is Kip S. Thorne, an associate professor of theoretical physics at CalTech, who is barely out of his twenties. Much of what follows is based on his work . . . and the printouts of his computer.

When a star loses the last of its nuclear fuel, or at least loses so much that gas and radiation pressure can no longer keep the star expanded, the ever-present gravitational force in the star becomes the dominant factor in its fate.

For stars of the Sun's mass, the story appears to be straightforward. Computer analyses tell us that once gas and radiation pressures can no longer support the star's size, gravity begins to compress the star. It falls inward on itself. The interior density and temperature rise as the gravitational collapse progresses, and eventually this produces a breaking action.

The Sun's eventual collapse may take place over the span of a few million years. Gradually it will sink from its grossly distended red-giant diameter to a diameter more like our own Earth's—about 12,700 kilometers—and its central temperature will reach nearly a billion degrees. The density at the core will go up to about a thousand tons per cubic inch. The Sun will be a white dwarf star.

Why does the gravitational collapse stop at this point? The star is composed of a plasma, which consists of ions (atomic nuclei that have been stripped of their orbital electrons) and the freed electrons. As the density of the plasma increases, these particles collide more and more frequently. The electrons, which can be thought of as a hazy cloud rather than a firm particle, can undergo some compression. And they do, getting squeezed further and further as the gravitational collapse forces the star's density higher and higher. At a density of about a thousand tons per cubic inch, though, the electrons resist further compression.

This produces the braking action—the counterforce that finally balances out against gravity and stops any further collapse of the star.

So now we have a star that's about the size of the Earth, although it still contains just about 2×10^{27} tons of matter.

During this contraction phase there may have been some unburnable fusible material in the Sun's outer layers. But as the interior temperature soared, any fusible elements, from hydrogen to iron, would eventually be heated to their ignition temperature and go off like a bomb. Thorne believes that this may explain the pulsars.

The final fate of the Sun after it has reached white dwarfdom seems rather prosaic. It simply cools off, as the heat generated from the collapse is slowly dissipated into space. The process may take billions of years, but eventually the Sun will be nothing more than a cold, dark body, the size of the Earth, with a density of some thousand tons per cubic inch.

But if it's drama you want, consider the fate of the more massive stars.

The computer calculations show that stars with more than 1.4 times the Sun's mass don't stop their gravitational collapse when they reach the white-dwarf stage. For stars this massive, the electrons' resistance to compression doesn't give enough of a braking force to counteract the gravitational force. The collapse goes on. There are a number of different possibilities as to what happens next. Much depends on the details of the individual star's mass, spin rate, and chemical composition. But the general outlines of the story appear to be firm.

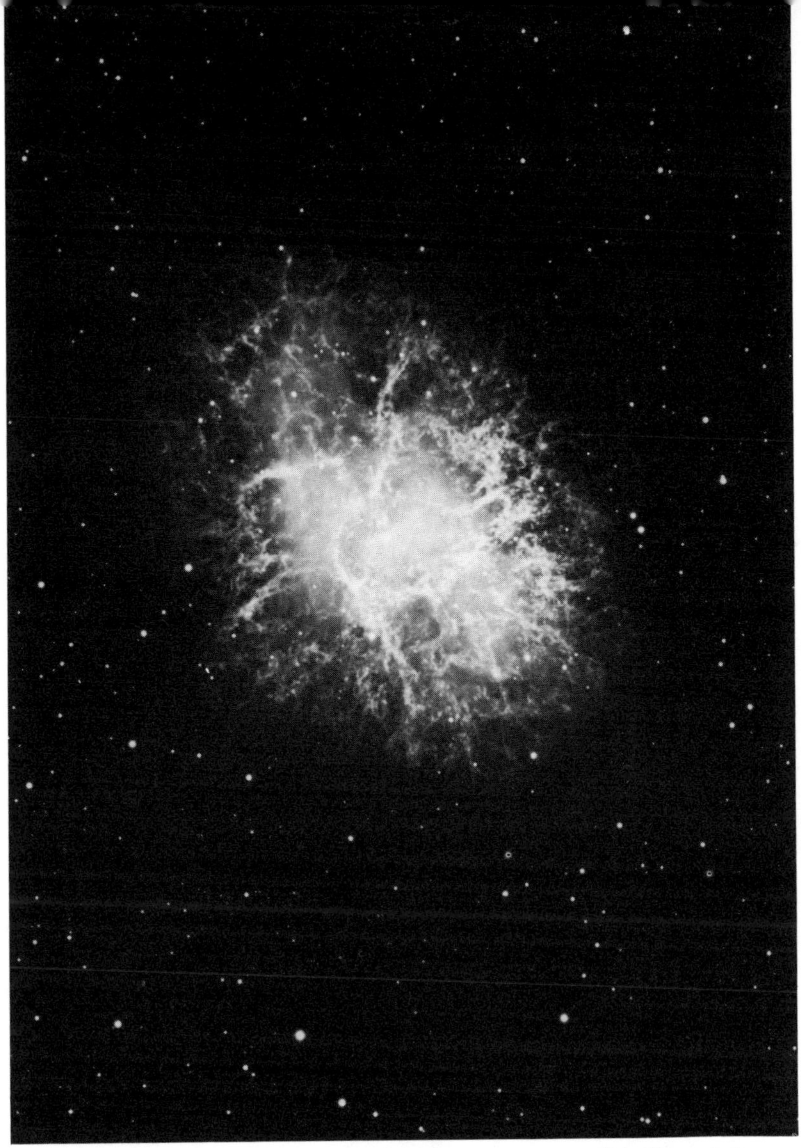

Hale Observatories

The results of a supernova explosion. The famous Crab nebula in the constellation Taurus. This nebula was seen as a supernova by Oriental astronomers on July 4, A.D. 1054. The distended cloud of plasma, still expanding at several hundred kilometers per second, emits radio and X-ray energy as well as visible light. At the center of the Crab nebula is a tiny pulsar, all that remains of the original star.

If there is enough unburned fusible material in the star's outer shell, the rapidly rising heat of the core may trigger a supernova explosion. As we have already seen, in a supernova the star may release as much as a billion years' worth of solar output inside of 24 hours. And although it seems hard to picture anything as surviving such a blast, it now seems certain that the core of the star remains relatively intact, at least for this type of supernova.

Whether or not there is a supernova explosion, the core of the star keeps on shrinking, past the density of a white dwarf. As the star's diameter keeps getting smaller and its density higher, gravity gets stronger and stronger. If the original star was massive enough, the gravitational force eventually becomes so powerful that the electrons can no longer resist further compression. They are squeezed into the atomic nuclei, turning all the protons in the nuclei into neutrons.

We now have a mass roughly equal to the Sun's, consisting entirely of neutrons, some 10^{57} of them, packed side by side in a sphere no more than 100 kilometers wide—probably more like 10 kilometers wide. Density is around a billion tons per cubic inch.

That's a neutron star.

If the star is not more than two solar masses, the tremendous repulsive forces that the neutrons exert on each other will resist any further gravitational crushing. The brakes are on—neutron brakes this time—and the collapse stops.

But the story does not end there. Far from it! The star's core has collapsed to neutron-star dimensions. But there are still outer layers of the star, even if much of this material has been blown off in one or more explosions.

This outer shell falls in on the tiny neutron core, since gravity is always hard at work. The impact creates enough heat to drive the core's surface temperature up to billions of degrees for a fraction of a second. Under these circumstances most of the heat energy is converted into neutrinos. Not to be confused with neutrons, neutrinos are aloof little particles, much like photons except that under ordinary circumstances a neutrino can penetrate 50 light-years of lead without being stopped.

But the conditions around a neutron star are far from ordinary. The densities and temperatures of the plasma around the neutron core are so high that even the evasive neutrinos can travel only a few meters before they are deflected or absorbed. Most of their enormous energy is imparted to the plasma clouds, heating them to tens of billions of degrees. This creates a supernova explosion, of course. But this is a different type of supernova—a *core supernova* that blows away everything except the tiny neutron-star core.

All this—the collapse into the neutron core, the infall of the shell of plasma, the heating that forms neutrinos, the core supernova explosion—all this happens in a few seconds.

The results? Look at the Crab nebula, that cloud of plasma a few light-years across, still expanding at several hundred kilometers per second more than 900 years after the core supernova that created it, emitting visible light, radio waves, X-rays, and even gamma radiation.

And in the center of the Crab nebula, beautifully verifying the whole theoretical story, is a pulsar!

For the pulsars, most astronomers firmly believe, are actually neutron stars that are emitting sharply timed bursts of radio energy.

The first pulsar discovered, CP 1919, is in the constellation Vulpecula, the Little Fox, a faint and shapeless star group that lies between the bright stars Vega and Altair. Shortly afterward, a half dozen additional pulsars were found. More are being detected constantly. The Crab nebula pulsar is designated NP 0532.

When the first pulsars were discovered, their precisely timed radio bursts led some astronomers to wonder if these signals might not be coming from an intelligent civilization in space. For a few weeks they were informally called LGM signals—for "little green men."

But by the end of 1967, Thomas Gold of Cornell and several other theoreticians had proposed natural models that seemed to explain the pulsar phenomenon very well. All of these models dealt with white dwarf or neutron stars. Using "Ockam's razor," the more complex explanation of an interstellar civilization was dropped in favor of the natural-phenomenon explanation.

Gold's explanation seems to be the most widely accepted at present. He pictures the pulsar as a neutron star surrounded by fairly dense plasma clouds (the remnants of the core supernova, probably), with the whole complex of core and clouds held together by a strong magnetic field. If the neutron star is rotating, which it no doubt would be, the rotation of its magnetic field would drag the plasma cloud around with it. However, the farther away from the star's surface you go, the faster the plasma must rotate to keep up with the forces pulling on it. This is like a "crack the whip" situation—tail-end Charlie must go like mad just to stay up with the rest of the gang.

At a far enough distance, the plasma simply cannot keep up, even though it may be moving at speeds close to the speed of light. Part of the plasma breaks away from the magnetic field, and in the relativistic processes involved, a beam of radio energy is formed. This happens on every rotation of the neutron star, causing a regular periodicity to the radio pulses. The observed timing of the pulsars' radio bursts, all grouped around the once-a-second mark, fit in well with the expected spin rate of a 10-kilometer-wide neutron star.

Thorne prefers a different explanation. You recall that white dwarf stars may produce explosions on their simmering surfaces—sort of super-flares? Thorne believes that surface flares or explosions on a rotating white dwarf might explain the pulsars. The flares would emit a beam of radio waves as well as visible light. But this theory does not seem to explain as well as Gold's the exactness of the pulsars' timing.

The discovery of the Crab nebula pulsar brought enormous support to Gold's explanation. Here is exactly the situation he postulated: a neutron star embedded in a plasma cloud laced with a strong magnetic field. Early in 1969 the Crab nebula pulsar was detected visually, photographed, and even scanned by television. The optical pulsations, in synchronization with the radio pulses, provided even further excitement and satisfaction for the astronomers.

Moreover, for a few of the pulsars, the periods of pulsation are *increasing.* This most likely means that they're still shrinking, still being crushed to smaller size by gravity.

How far can the crush go?

The answer, according to theoretical physicists such as Thorne, is that under right circumstances a star can literally go straight out of this universe.

Let's take another look at a neutron star. With our eyes of theory we can see through the swirling carnage of plasma that surround the star.

Originally a star of more than 1.4 solar masses, it has suffered a gravitational collapse, perhaps gone through either an outer envelope or core supernova (or both?) and now is reduced to a neutron core with a mass between one fifth and twice the Sun's mass. For several thousand years such a neutron star will emit more X-ray energy than the Sun's output of visible light. And as we've seen, it is also causing radio and optical pulses, although these are most likely coming from the surrounding plasma clouds and not from the neutron star itself. After a hundred million years or so, the neutron star's temperature will cool down to a few thousand degrees, and it will become a quiet, dark chunk of matter some 10 to 100 kilometers across.

But if the neutron core of a collapsing star is more than twice the Sun's mass, its gravitational infall won't stop at the neutron-star stage. It will keep on shrinking. As the interior density goes past the 10-billion-tons-per-cubic-inch mark, the neutrons themselves are squeezed down into smaller particles called hyperons. Classical physics can't describe what happens now—only relativistic physics can.

To paraphrase an old joke, the star digs a hole, jumps in, and pulls the hole in after it!

For, once the gravitational collapse goes past the neutron-star stage, the star is on a one-way ride to total oblivion. It will disappear from this universe.

If the star's neutron core is more than twice the Sun's mass, no possible braking force can stop the collapse. J. Robert Oppenheimer began studying this kind of ultimate gravitational collapse back in 1939, together with Hartland Snyder of the University of California at Berkeley. Oppenheimer was soon diverted into the Manhattan Project and later into the ultimate tragedy of the Joseph McCarthy era. He never returned to this particular aspect of physics again. In fact, it wasn't until the mid-1960's, when the work on quasars and on supernova explosions both led to the problem of gravitational collapse, that the subject was really reopened for further examination.

As the collapsing star squeezes in on itself, compressing the same amount of matter into an ever-smaller radius, the gravitational field of the star becomes titanic. Photons emitted by the star must work against the gravitational field to escape the star's vicinity. We saw earlier that a body of 10^8 solar masses would have a gravitational field strong enough to produce large red shifts in the photons emitted.

Now we're talking about single stars, whose mass is comparable to that of the Sun. But if the Sun shrank down to a diameter of 5.8 kilometers, its gravitational field would be so strong that no photons could escape its surface. The Sun would disappear.

For a body of the Sun's mass, the *gravitational radius* (the radius at which photons can no longer escape) is 2.9 kilometers. But the Sun will never shrink that far, so the computer runs tell us.

For a star with a neutron core of more than two solar masses, though, a black pit is waiting.

The star reaches neutron-star density but keeps right on shrinking—gravity is so powerful that it overrides everything else. When the star's diameter gets down to about 6 kilometers, it winks out. Photons can't escape from it any longer. It has dug a black hole in space and disappeared into it.

It's a strange place, this black hole. The gravitational collapse doesn't stop simply because we can't see the

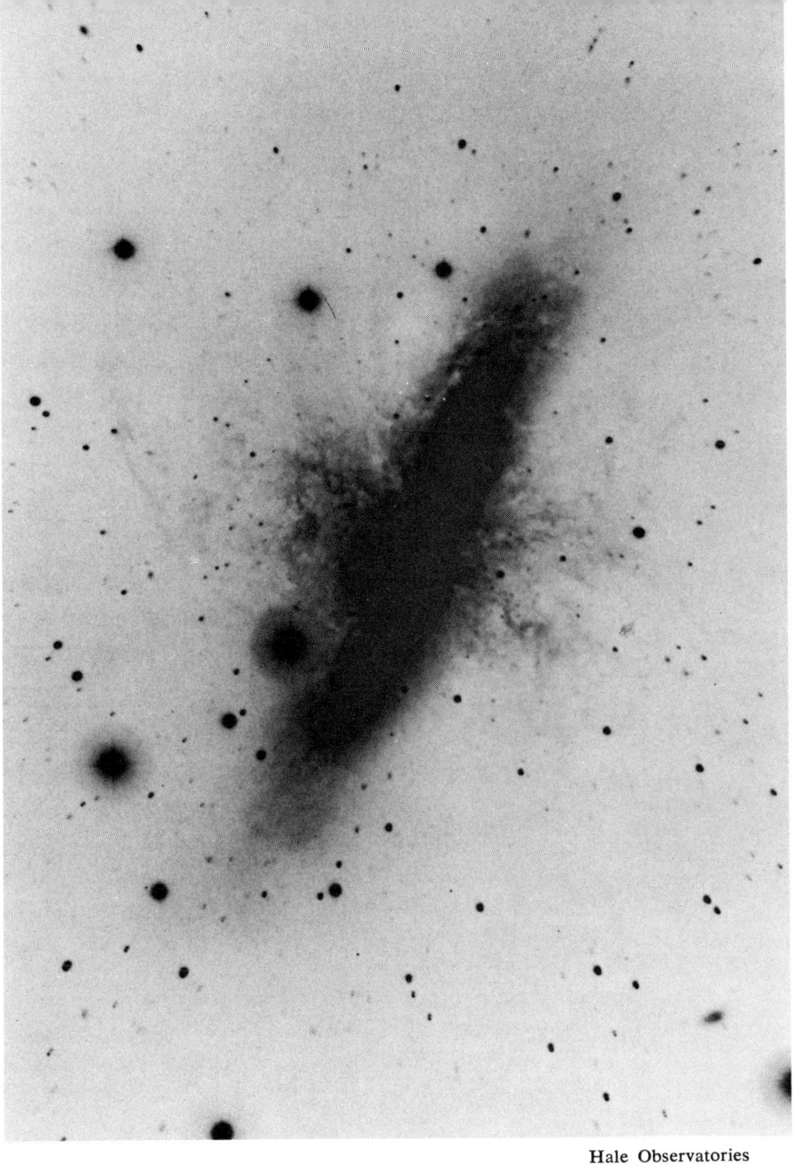

Hale Observatories

An exploding galaxy. This is a negative, easier to use in some cases than a positive print, because it is easier to trace black on white. The small galaxy M 82 is hurling filaments of plasma outward from its core for distances of some 25,000 light-years. The cause of the explosion is unknown, but astronomers now believe that many galaxies—including our own Milky Way—suffer core explosions.

star anymore. According to theory, the star keeps collapsing until it reaches zero volume and infinite density! Such a point is called a *Schwarzschild singularity,* after the German physicist Karl Schwarzschild (1873–1916).

More on the singularity aspect in a moment.

We have watched a star collapse into the white-dwarf stage, explode, and collapse into a neutron star. What would it look like if we could watch the final disappearance of a star as it collapsed down into a black hole?

First, we would have to be able to see through the plasma clouds that surround the scene. Second, we would have to be able to see in X-ray wavelengths, because that's what the star is now radiating. Finally, we would have to look impossibly fast, for the whole thing happens in less than a second. (Under the influence of such titanic gravitational fields, however, time itself may become warped out of all recognition.)

The star visibly collapses and becomes "redder" as it shrinks. The photons must work harder and harder to get away from that fast-increasing gravitational field. Perhaps they are even shifted to visible wavelengths. But assuming you can see the star, regardless of the wavelength of its radiation, one moment it's hanging there in the midst of its plasma nebulosity, then suddenly it shrinks like a pricked balloon, getting smaller and smaller.

At last—still within the space of a second, remember—the collapse will seem to slow down. A few photons are struggling up out of the rim of that black hole. The star finally disappears, but there's a dim halo left, a few kilometers across, where those last few photons are taking tortuously spiraling paths to work out of the gravitational pit that the star has dug for itself.

The planets around such a star should be safe from falling into the gravitational pit, although any native life on such worlds would have been scoured away during the earlier explosions. We, in a spacecraft, can approach the star quite closely, as close as the gravitational radius itself, without being sucked into the black hole. The chances of losing interstellar spacecraft due to stray wanderings into invisible black holes are much smaller than the chances of being punctured by a millimeter-sized meteoroid in our own solar system.

Or are they?

Gravitational collapse down into a black hole can also happen for objects larger than individual stars—for whole galaxies or quasars, in fact. The gravitational radius for a galaxy of a billion solar masses would be roughly one fifth of an astronomical unit. With a diameter of some 15 million kilometers, you could fit several collapsed galaxies inside the orbit of Mercury! The gravitational effects on the rest of the solar system might be interesting—if gravitational waves from the collapsed galaxy can escape the black hole.

Even a "pothole" 0.2 Astronomical Units wide is microscopically small in the superhighway of interstellar space. But if a craft should ever hit one, it will disappear forever.

Or will it?

In that strange world inside the black hole, where a star is crushed down to zero volume and infinite density, what physical rules apply? Even relativistic physics has nothing to say when the density gets to something like 10^{87} tons per cubic inch. This is 10^{78} times denser than a neutron star. If the Sun were made that dense, its size would be about one millionth the diameter *of an atomic nucleus.* Nobody knows what happens, physically, down at the bottom of the black pit.

Except that it might be bottomless. Or better yet, open-ended.

Several theoreticians have pointed out that the mathematics of gravitational collapse and Schwarzschild singularities apply only to perfectly spherical bodies. Stars are not perfect spheres, and certainly galaxies are even less so. As they're gravitationally crushed, it's likely that any deformations in their shapes will become exaggerated, not smoothed out. Roger Penrose, an English mathematical physicist, has shown that a nonspherical body may collapse down toward a singularity, but would not be completely crushed to zero volume. For reasons known only to the mathematicians, such a body can escape going to a singularity. But it can't stay in the same physical location where it collapsed. In effect, the black hole turns into a tunnel.

You can visualize this by drawing on an analogy that relativistic physicists have often used. Picture spacetime as being represented by a thin, very flexible sheet of rubber. We'll picture it as a flat sheet, although actually it's probably curved and may be quite intricately convoluted. Massive bodies such as stars can then be thought of as tiny ball bearings resting on this rubber sheet. The bigger and heavier the star, the deeper the dimple it makes in the otherwise smooth sheet.

For a star or galaxy that's collapsing into a black hole, this dimple starts to look more like a tunnel—a long, thin tube stretched in the fabric of space-time by the gravitational collapse of a massive body.

If the body does not go down to a singularity, then the tube-tunnel might emerge *somewhere else in space-time.* The star or galaxy has dug its way out of one place in the universe and reappeared somewhere, and perhaps sometime, else.

No one has seriously proposed explaining the physics of this phenomenon. Where even relativistic physics breaks down, you can't expect more than a shrug of the shoulders when you ask questions. Perhaps the enormous energy locked in the star's gravitational field is the driving force behind its tunnel-drilling. Certainly, at the densities and gravitational-field strengths involved, it seems clear that the entire fabric of space-time gets badly bent. Dare one use the science-fiction term of "space warp"?

Some cosmologists have seized on this idea to propose that the quasars are themselves the explosive reemergence of collapsed galaxies, bursting back into our universe with a loud, bright bang after having tunneled their way out of black pits.

Maybe.

But what are those tunnels like? Do they stay intact, a sort of underground railway system crisscrossing the fabric of space? Could a spacecraft take a shortcut, and maybe break the universal speed limit by going through a tunnel? What happens to *time* inside such a tunnel?

Ultimately, if the universe is finite, its expansion will slow down, stop, and the final gravitational contraction will begin. Will the universe end in a black pit? Or will we, all squeezed down to hyperons at least, tunnel out into a new and different universe?

Whichever way it goes, the answer is a long time in the future. For the present, it's fun to consider the possibility of tunnels through space. Far from avoiding black holes, someday our spacecraft may be seeking them out, looking for the Northwest Passage between here and the Clouds of Magellan.